THE FIFTEEN SCHOOLGIRLS

THE FIFTEEN SCHOOLGIRLS

a brief account of a mathematical problem

and its originator

DICK TAHTA

First published in Great Britain by Black Apollo Press, 2006

Copyright © Dick Tahta 2006

The moral right of the author has been asserted.

A CIP catalogue record of this book is available at the British Library.

ISBN: 1 900355 48 5

Design: DCG DESIGN, Cambridge

Contents

Les Grecs se plaisaient à donner à presque tous les jeux des jeunes filles une forme orchestrique. Quand une fête s'approchait, ne fallait-il pas préparer les cérémonies, répéter les chants, s'instruire dans les évolutions du choeur? ...

C'est ainsi que l'on pourrait considérer comme jeux de jeunes filles la plupart des danses qui nous sont décrites par les auteurs anciens. Les rondes enfantines, les colliers ou rondes alternées, les promenades des jeunes filles par deux ou par trois donnent lieu à un certain nombre de problème nouveaux et intéressants qui se rapporter à la théorie des combinaisons.

(Edouard Lucas)

Every collection of units has a definite form, due to the number of its component units, and to the way in which the distinguished and undistinguished units, pairs, triads, &c., are distinguished throughout the collection. ...

Mathematics is the science by which we investigate those characteristics of any subject-matter of thought which are due to the conception that it consists of a number of differing and non-differing individuals and pluralities.

(Alfred Kempe)

All things began in order, so shall they end, and so shall they begin again; according to the ordainer of order and mystical mathematics of the city of heaven.

(Thomas Browne)

Prologue

The Lady's and Gentleman's Diary was an annual publication founded in 1841 as an amalgamation of two separate Diaries dating from the eighteenth century. It was the custom to include some mathematical puzzles and the Diary for 1850 included the following:

> VI Query: *by the* Rev. THOS. P. KIRKMAN, *Croft, near Warrington.* Fifteen young ladies in a school walked out three abreast for seven days in succession: it is required to arrange them daily, so that no two shall walk twice abreast.

This was to become a famous and much explored problem that engaged a number of people at the time and since. It also turned out to be part of a more general combinatorial problem which had a mathematical significance that was only gradually understood. The following pages attempt to explore some of the mathematics involved and some of the life and work of the proposer of the problem.

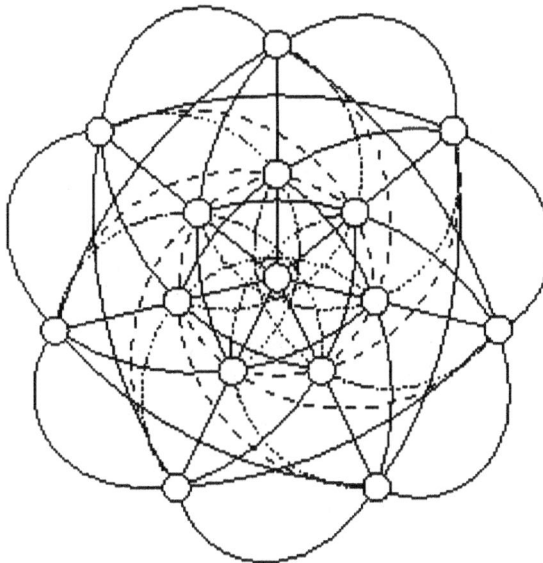

Kirkman and his first mathematical paper

1.1 Thomas Penyngton Kirkman

Kirkman was born in 1806 in Bolton, Lancashire. His father was a cotton merchant who wanted his only son to join the family business. Thomas went to Bolton Grammar School, but left unwillingly at the age of fourteen, to work with his father for the next nine years. This must have been a difficult time for him, but he continued to study on his own and he finally rebelled and enrolled at the age of 23 as a student at Trinity College, Dublin. He paid for himself by taking on some tutoring and after he graduated in 1833 he stayed in Ireland for another year as a private tutor.

On his return to England in 1834, he joined the church and obtained a curate's post at Bury, though he was not in fact ordained until two years later, when he moved to be a curate in Lymm, Cheshire. Then in 1839, at the age of 33, he moved to Croft, a village near Warrington. This turning point in his life and what he made of it was later described by his eldest son, William, in an obituary:

> He was enticed by fair words, by the then rector of Winwick, to bury himself for life as the rector of the newly-formed Parish of Southworth with Croft, where he remained for 52 years. Here, by perseverance and his gift of teaching, he formed, out of the roughest material, a parish choir of boys and girls who could sing at sight any four-part song set before them. Here also, with an expenditure of mental labour that only the finest of physical constitutions could have sustained, he devoted, practically, the whole of his time (for the parochial work was small) to the study of pure mathematics, the higher criticism of the Old Testament, and questions of first principles.

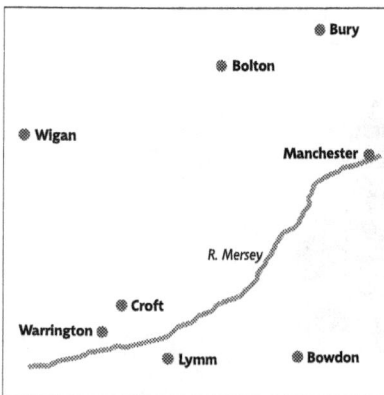

Kirkman married Eliza Wright in 1841 and they were to have seven children (five sons and two daughters) over the next thirteen years.

In the circumstances it may seem surprising that in his fortieth year, 1846, he presented a mathematical paper to a well-known Literary and Philosophical Society in Manchester. This was the first indication that he had any specific mathematical interests, but it was to be the first of a series of articles he published for the rest of his life. His paper involved a combinatorial problem which arose from a puzzle that had been circulated in an annual publication some years before. He submitted his own puzzle about the fifteen schoolgirls, which was a particular case of the problem he had been working on, a few years later in 1850.

He published a number of further mathematical papers on combinatorial problems over the next few years in specialist mathematical journals, and corresponded with many of the leading mathematicians of the time. He also worked at some related geometrical issues, including a study of the arrangements of Pascal lines associated with a hexagon inscribed in a conic, and some lengthy papers on the classification of certain polyhedra. The latter topic seems to have been stimulated by a Prize Question proposed by the French Académie des Sciences in 1855. There had been no response to this challenge and the prize had been held over for a few years. Kirkman wrote a lengthy memoir which he submitted to the Royal Society which published part of it; he had been appointed a Fellow in 1857.

He had not responded directly to the Prize Question on polyhedra because his attention had been caught by another question proposed for the prize which was to be awarded in 1860. This was on an algebraic problem on what would now be called permutation groups. Kirkman prepared and submitted a lengthy memoir on the subject, written in French over a period of two years in which he was also writing his even more lengthy paper on polyhedra. But in the event the prize was not awarded. Kirkman was disappointed and became quite obsessional about the failings of the Académie. This paralleled his feelings about what he felt were the failings of the Royal Society in not publishing all of his work on polyhedra.

He continued to work on various topics, but began to feel somewhat isolated from other mathematicians. He had fallen out with Cayley whom he blamed for the neglect of his work on polyhedra. He also fell out with Sylvester over an argument about priority in work associated with the puzzle about the fifteen schoolgirls. At one stage he thought of trying to get some appointment at Cambridge and enlisted the help of de Morgan who wrote on his behalf to Trinity College. Part of this letter is quoted in an account of Kirkman's life and work by Norman Biggs:

> He is buried at Croft and very much desires a better field of action in which he may be able to see a little more of intellectual life, over and above his clerical doings. He has barely £180 a year, and is an active man, moderate and of orthodox repute ... He has worked for many years at subjects which will not bring him before the general eye, and is a staunch enthusiast and *con amore* mathematician.

Nothing came of this. Kirkman's mathematical work remained relatively unknown. According to Biggs, "It must have been convenient for the mathematical establishment to look upon him as a crazy cleric who had invented an amusing puzzle about schoolgirls. And it must be said that he himself did little to dispel this illusion."

Apart from his mathematical work, Kirkman wrote on other issues that he felt passionately about. One early book that was published in 1852 was called *First mnemonic lessons in geometry, algebra and trigonometry*. This arose from his private tutoring of young schoolchildren in Croft and offered some curious condensed phrases that were supposed to make it easier to recall certain mathematical rules. Thus Pythagoras' theorem was to be recalled as 'qua poth is both qua sides' where qua(drate) = square of, and poth = (hy)poth(eneuse). The cosine formula was coded as 'sq b is DUQ ac le CoBang two ac' where sq b = square of b, le = 'less', DUQ = duo quadrata = sum of squares of, and CoBang = cosine of B angle; this mnemonic was meant to be read as 'squib is duck ac le cobang two ac'.

There were lots of these mnemonics introduced in the course of discussions between 'Uncle Penyington' and two fictional children, Richard and Jane. (These have always been favourite names for educational writers – some readers may recall their re-incarnation in Ladybird reading schemes.)

> **Do you understand all this?**
>
> *Jane:*—I think I see that every step of the argument is proved; but I know not where I am, or what is before me, and cannot see much of what is behind me. It is like plunging into a dark cavern guided by a slender thread: I have just hold of it, and that is all.
>
> *Uncle Pen.:*—It will never break, for the twine is indestructible; and there will be light enough presently. If you are convinced that equals divided by equals give equal quotients, you are certain that Bb and Cc are true; and if equals multiplied by the same quantity remain still equals, B', b', C', c', are true likewise. When you see all the meaning and application of these results, you will know that they contain the whole science of geometry.

One aside to the girl suggests that Kirkman did not think mathematics was a suitable subject for women: "Take some pains to master them, my dear Jane, and you will save your time, of which you have very little to spare for mathematics, a study in which I should be sorry to see you deeply engaged". On the other hand, he gives her some interesting insights; for example, her comment on imaginaries: "So whatever mystery or appearance of contradiction there may be here, it springs not from the answer or the oracle, but from the ignorance of the interrogator. His duty is not to cavil at the response, but to go away ashamed of himself and wondering". Moreover, Jane is shown as understanding the lesson when the teacher suggests that

Richard has been finding it rather difficult.

Richard: Indeed I do, and have left to Jane the honour of following you all through this lesson. How do you feel, dear sister of mine, after this feast of trigonometry?

Jane: A little fatigued; but more by the variety than the difficulty of the subject.

More interesting perhaps was his introductory explanation of the proposed method of making mathematics more accessible:

If the experience of others in tuition agrees with my own, I may perhaps look to reap a little praise – not mathematical, on ground like this, but simply didactical – the praise of teaching well, of which I confess myself covetous. It appears to me that distaste for mathematical study often springs, not so much from any abstruseness in the subject at any point, to the student who has mastered the approaches, as from the difficulty generally felt in retaining the previous results and reasoning. This difficulty is closely connected with the *unpronounceableness of formulae*: the memory of the tongue and of the ear are not easily turned to account; nearly everything depends on the thinking faculty, or on the practice of the eye alone … My object is to enable the learner to *talk to himself* in rapid, rigorous and suggestive syllables, about the matters which he must digest and remember.

There cannot have been many sales of this extraordinary book. The author himself doubted "that any suggestion of his, for the improvement of established methods of tuition, will be favourably received, or confessed to have any value".

Kirkman also wrote on other than mathematical issues. He was critical of the orthodox church opinion that had been shocked by the views of John Colenso, Bishop of Natal, who had suggested, in 1862, that the Pentateuch, the first five books of the Old Testament, were not literally true. He gave two public lectures in Croft in defense of the Bishop and these were published as pamphlets in 1865. In the first, *Truth against tradition*, he wrote: "I believe that it is a dangerous error to say that every word, every syllable, every letter of the Bible is just what it would be if God had spoken it from heaven to us without human intervention". Later he was to feel that his stance had affected his chances of promotion in the Church.

He was a fanatical opponent of what he called materialist philosophers who were, he felt, peddling dangerous evolutionary ideas. He wrote more than a dozen pamphlets over two decades. He was particularly critical of

---

An extract from the 1865 lecture by Kirkman: *Truth against tradition*

...Truth has conquered against tradition. No man in Europe of the smallest education now believes that the earth stands still, and that the sun moves round it. Tradition is confessed to be erroneous, in spite of the letter of the Old Testament, in spite of all nations in all past times, and in spite of the deep tones of solemn doctors, who are still training their young sons of thunder to affirm, with might and main, that "modern science, with all its wonderful advances, has discovered not one single inaccurate allusion to physical truth in all the countless illustrations employed in the Bible".

It may easily be, that many of the bishops and dignitaries, who have been lately clamouring for punishment on the good Bishop of Natal, had great-grandfathers living when Galileo was trampled on ...

Herbert Spencer whom he accused of talking nonsense by dealing with "abstracts in the clouds, instead of building on the witness of his own self-consciousness". He made fun of Spencer's definition of evolution ("a change from an indefinite incoherent homogeneity to a definite heterogeneity ...") by writing that, "Evolution is a change from a nohowish untalkaboutable all-likeness, to a somehowish and in-general-talkaboutable not-all-likeness, by continuous somethingelsifications and sticktogetherness".

He published more than a dozen short pamphlets in his sixties, a decade when he does not seem to have been so involved with mathematics as he had been. These pamphlets were on various issues, with titles such as 'The ordeal of jealousy', 'Philosophy without assumptions', 'Clerical dishonesty' and 'An address at the twenty-first soiree of the Brighouse Mechanics' Institute'.

Though he was never to be again so prolific as in his first two decades at

---

An extract from an 1887 article by Kirkman: *On Mr. Herbert Spencer's conquest of the problem of the universe.*

Mr Spencer has many things to say about the infinite, all wonderfully old *testimonia paupertatis*. Meanwhile, men of accurate thought, whether mathematicians or not, go on satisfying themselves that, when for answer to a clear question that all agree will be asked, it is evidently absurd to accept either nothing or finite, we are justified in recording infinite as the true reply. We do not pretend to conceive the maximum final term of the series of natural numbers 1,2,34,&c.; but when we say it is infinite and nothing less, we know that the man is an idiot who replies – 'Yes, but what is on the other side of your infinite?' In the same way, when from the facts of the present we draw any other given line of thought that loses itself in the infinite, and have agreed to call the reality at that further end First or Ultimate, it takes a couple of idiots rolled into one to frame the questions – How came there to be a First? How came that Ultimate there? Yet the fact that given elements of reasoned thought lead us on towards the infinite does not make us all eager, like nimble Mr Spencer, to gallop away from the field of clear question and answer about what is close to us, right up to infinite at once, and to begin our arithmetic or philosophy there – still less our religion.

Croft, he continued to work and publish papers on the various mathematical topics he had been interested in, as well as embarking on new ones, such as the theory of knots in which he published papers in his eighties. He had been introduced to this topic by the mathematician, Peter Tait, who shared his views of Spencer. He also continued to submit numerous problems and solutions to a half-yearly collection of mathematical puzzles after his retiremen in 1892 when he moved to Bowdon. These included two problems in the theory of equations published in the year of his death in 1895. His wife Eliza died twelve days later.

A final remark is taken from one of Kirkman's letters to a correspondent: "What I have done in helping busy Tait in knots is, like the much more difficult and extensive things I have done in polyhedra or groups, not at all likely to be talked about intelligently by people as long as I live. But it is a faint pleasure to think it will one day win a little praise."

1.2 A problem in combinations

Kirkman became a member of the Manchester Literary and Philosophical Society soon after he moved to Croft. At a meeting of the society in December, 1846, he read a paper, *On a problem in combinations*. This was printed in the *Cambridge and Dublin Mathematical Journal* the following year; it was his first publication. In his paper, Kirkman introduced his combinatorial problem as a particular case of a more general one, namely to find

> ### On a Problem in Combinations (read before the Literary and Philosophical Society of Manchester, 15 December 1846)
>
> If $Q(x,y,z)$ denote the greatest number of combinations of y together, that can be made with x symbols, so that no combination of z together be twice employed; $Q(x,y,1)$ is the greatest integer in x/y and $Q(x,y,y)$ is the number of combinations of y together that can be made with x things. Thus division of integers, and this simple problem of combinations, would appear to be particular cases of the more general problem, whose solution is $Q(x,y,z)$.
>
> The object of this paper is to assign $Q(x,3,2)$; and to establish the following theorem:
>
> If $Q(x)$ denote the greatest number of triads that can be formed with x symbols, so that no duad shall be twice employed, then $3Q(x) = x(x-1)/2 - V(x)$, if for $V(x)$ we put $6k+4$ when $x=6n-1$; $x/2+3k+1$ when $x=6n-2$; and $x/2$ when $x=6n$ or $6n+2$: where $2^m.(2k+1)=n$; x,n,m,k, being all integers ≥ 0.
>
> ....

"the greatest number of combinations of y together, that can be made with x symbols, so that no combination of z together shall be twice employed".

Such a collection may be referred to as a (y,z) system for the x elements. Thus, with $z=1$, a $(y,1)$ system will contain the greatest number of disjoint sets of y elements that can be formed from x elements, and the number of sets in this case is the greatest integer in y/x. With $z=2$, the simplest case is when $y=2$, and the problem is to find the number of distinct pairs that can be formed from x elements; for example, the intersections of a given number of lines, or the possible handshakes among a group of people.

The next simplest case is that of a $(3,2)$ system and this is the particular problem investigated by Kirkman, namely to select from x elements the greatest number of triples with no repeated pairs. Some examples of such systems are as follows:

- How many triangles without any sides in common can you make on a 9-nail geoboard?
- You have 7 friends whom you wish to invite to dinner in threes. How many times can you do this before two of them come together for a second time?
- How many 3-letter English words can you make with 10 letters, with no pair of letters appearing twice?
- The problem about fifteen schoolgirls asks how many triples with no repeated pairs can be made from 15 elements.

How had Kirkman become interested in his combinatorial problem? There is no record of his having any special mathematical interests before 1846. But the particular problem about triples with unrepeated pairs had appeared in the *Lady's and Gentleman's Diary* for that year. The *Diary* was an annual publication which regularly included a mathematical Prize Question with solutions from readers being printed in the following year. The problem about triples was carried over from the more general Question that had been set two years previously, in 1844, by the editor of the *Diary*, W.S.B.Woolhouse:

> Determine the number of combinations that can be made of n
> symbols, p symbols in each, with this limitation, that no combination
> of q symbols which may appear in any one of them shall be repeated
> in any other.

This may be described as seeking a (p,q) system for n elements. Some responses from readers were printed in the *Diary* for 1845. The editor pointed out that a solution from Septimus Tebay from Preston assumed that every possible combination of q elements was included. In the following year the question was repeated, but only for the special case $p=3$ and $q=2$.

The editor pointed out again that with, say, 10 elements it would not be possible to cover all pairs. Thus, with 10 elements there will be 9 pairs containing any particular element. But these will appear two at a time in any triple including that element, so no set of triples could contain all the 9 pairs without a repeat of one of them. A similar argument shows that if there is to be a *complete triple system*, namely one of triples that include all possible pairs without repeats, then n must be odd.

It seems that Kirkman began his combinatorial investigation in response to a challenge in a popular annual publication. This was certainly what he was to recall some years later. The question of how the problem arose for Woolhouse himself will be taken up later.

The main result in Kirkman's paper concerned what has here been called complete triple systems. As noted above, all possible pairs cannot be achieved when x (the number of elements in Kirkman's notation), which will be retained) is even; for a complete system x must be odd. Moreover, since every triple yields 3 pairs the number of pairs, namely $x(x-1)/2$, must be divisible by 3. Hence, x must be of the form $6k+1$ or $6k+3$. Are these conditions sufficient – that is, will there be a complete triple system for all values of x of the necessary form? Kirkman proved that there was by showing that if there was a complete system for x elements, then it was possible to construct a complete system for $2x+1$, and also for $4x-3$, elements. Since there is a complete system for 3 elements, there will also be systems for 7 and 9. Repeating this argument in turn proves the result for 13 and 15, and in general for all elements of the form $6k+1$, $6k+3$.

Kirkman showed further that when x is of the form $6k$ or $6k+2$, any triple system would have at least $x/2$ missing pairs. The situation for other values of x, such as 10,11,16,17 and so on, was more complicated. Kirkman constructed some systems for these cases, but found it was difficult to prove that they were the best possible. "The results about to be offered," he somewhat ambiguously wrote, "although they may perhaps be assented to as certain, will yet be found deficient in mathematical rigour."

Kirkman's complete triple systems were to be assiduously investigated and fruitfully generalised by many other mathematicians, in ways that will be described in later pages. But there was very little study of incomplete triple systems for some time. Recent work does not refer to Kirkman's 1846/7 paper which has been generally neglected except for occasional references to his results for complete systems.

It turns out that Kirkman was mistaken in some of the cases he was doubtful about. In 1966, an Israeli mathematician, Johann Schonheim, established some general results from which it follows that for x of the form $6k+4$ there will always be at least $x/2+1$ missing pairs, while for x of the form

$6k+5$ there are always at least 4. This means that for 16 elements, Kirkman proposed lower limit of 12 missing pairs can be reduced to 9, and that for 17 elements his lower limit of 10 can be reduced to 4. These corrections yield an elegant overall formula for the largest number of triples in a (3,2) system for x elements, namely $[x/3.[(x-1)/2]]$ – where the square brackets denote the greatest integer function – with 1 less than this in the case when x is of the form $6k+5$. This would surely have delighted Kirkman, whose formula was not only – with hindsight – incorrect, but also particularly cumbrous.

Meanwhile, the more general problem originally posed in the *Diary* has proved to be more difficult. Some years later, Kirkman did some further work on the problem of finding complete systems for multiples with no repeated certain submultiples; for example, (4,3) systems of quadruples with no repeated triples. He also investigated triple systems in which the pairs were in fact repeated but always with same frequency. The general problem was also taken up by others and developed in various ways; but the corresponding incomplete systems were again not studied until relatively recently.

It may be that one reason for the relative neglect of Kirkman's pioneering paper has been the popularity of his far more widely known presentation of a triple system in terms of schoolgirls. This puzzle was clearly a further development of his combinatorial researches. In fact, Kirkman later described how it was after the publication of his first paper that he noticed that there was a (3,2) system for 15 elements that could be resolved into 7 sets of 5 triples such that each element only occurred once in each of the sets. In other words, the (3,2) system could be resolved into seven (3,1) systems. It was this sevenfold resolution that allowed the convenient posing of the schoolgirls problem in terms of walks over a week. The general problem of resolving (3,2) systems into sets of (3,1) systems was taken up by many professional, as well as amateur, mathematicians, and was always associated with Kirkman. But, as will be seen later, triple systems as such were soon to be associated with another mathematician's name.

Before looking at such further developments, it is worth looking back to the genesis of the combinatorial problem which sparked Kirkman's interest and his subsequent mathematical career. Who was Woolhouse and how would he have come across the problem which he published in the *Lady's and Gentleman's Diary* for 1844?

1.3 Woolhouse and the *Diary* question

Westley Woolhouse was born in 1809 at North Shields. Three years younger than Kirkman, he also lived into his eighties. He seems to have developed a precocious ability in mathematics as a young boy: it said that he chalked up some difficult integrations on the shutters of his father's shop. By the age of thirteen, he was solving mathematical problems in various journals, and he published a book on geometry when he was nineteen. He was appointed to work on the *Nautical Almanack*. Some years later, he became a government actuary.

He seems to have had a thirst for knowledge and a breadth of interest that was common among professional scientists in the nineteenth century, especially those that were, like him, self-taught. He wrote books on musical intervals, on mortality in the Indian Army, and on the differential calculus, as well as some well-known tables of weights and measures. He later contributed to the journal of the College of Preceptors, *The Educational Times*, when it began a column of mathematical problems in 1862. He published various mathematical articles in learned journals on such diverse topics as vanishing fractions (whose numerator and denominator tend to zero), the theory of gradients in railways, and the numerical solution of equations. He was also the first editor of the annual *Lady's and Gentleman's Diary* which was founded in 1841 as an amalgamation of two diaries dating back to the eighteeenth century. The merger earned the scorn of one contributor, Noah Wilmot:

> Though unions are now all the rage,
> Who would have thought Lady Di
> Would at her patriarchal age
> Have listened to a lover's sigh?

But the amalgamated *Diary* was very successful, and Kirkman was particularly fulsome in its praise: "An incomparably greater share of the glory of kindling and cherishing a pure and lasting love of mathematics … must be attributed to the immortal Lady Di, than to all the universities and colleges".

Woolhouse – like many statisticians, then and now – sought to apply his mathematical skills to industrial problems. During the 1830s, there was a steadily growing reform movement to restrict the working hours of young people under eighteen and of all women. It was some years before a bill was passed restricting work to even ten hours a day. In the preparation of this bill, Woolhouse was asked to find out about working conditions in cotton factories. He estimated that in running to and fro attending to the spinning mules, the workers covered more than thirty miles a day.

It was obviously not the daily 'walks' of the young Lancashire cotton girls that Kirkman had in mind when he gave a popular presentation of Woolhouse's combinatorial problem. The contrast between the life of cotton workers and schoolgirls is an example of the sort of moral ambiguity that often occurs in the cover stories that people sometimes invent to make mathematics supposedly more palatable. But Woolhouse made no attempt to provide a setting when he set his combinatorial problem in the *Diary* for 1844, and he never offered any explanation later of how and why he came across it. We can only marvel at the appetite of his readers who were expected to get involved in a problem that was presented in such an abrupt and abstract way.

It is clear that he kept up with Kirkman's subsequent work, for he wrote various articles himself on the 15 schoolgirls. In 1861, he described how he had been prompted to write after he had read some historical remarks about the problem in the same journal earlier that year:

> On a recent perusal of this interesting paper, I could not help noticing the summary character of these allusions [to those mathematicians who had contributed to the problem] which first suggested to my mind the propriety of making the present communication with the view of pointing out the fact that the Rev. T. P. Kirkman originated this particular problem and that it first appeared in the *Diary*.

This refers to the first appearance of the 15 schoolgirls in 1850. Woolhouse began his article by quoting the general problem he had posed in 1844. He referred respectfully to Kirkman's investigations of the special case of triples and to the presentation of the schoolgirls problem. He then gave an account of some solutions, by readers of the *Diary*, that were due to be printed in the issue for the following year. He ended by offering a further analogous problem:

> Sixteen symbols may be arranged five times in the form of a square so that every pair of symbols shall appear once and only once both in a horizontal and a vertical line.

This appears to be the first example in the literature of a system of quadruples with no repeated pairs.

Woolhouse died in 1893 at the age of 83. An obituarist referred to his character as one of 'unblemished simplicity' and quoted a grandson who had remarked, somewhat ambiguously as it may appear now to us, that he was "a perfect child in business affairs, always too ready to please and too willing to be led".

The article that had upset Woolhouse in 1861 had been written by the eloquent and exuberant mathematician James Sylvester, who was in no doubt that he was the originator of the combinatorial problem that Kirkman had expressed in terms of schoolgirls.

> I may also take occasion to observe that in connexion with my researches in combinatorial aggregation, long before the publication of my unfinished paper in the Magazine, I had fallen upon the question of forming the heptadic aggregate of triadic synthemes comprising all the duad, to the base 15, which have since become well-known, and fluttered so many a gentle bosom, under the title of the fifteen schoolgirls problem, and it is not improbable that the question, under its existing form, may have orginated through channels which can no longer be traced in the oral communications made by myself to my fellow undergraduates at the University of Cambridge, long years before its first appearance, which I believe was in the *Lady's Diary* for some year which my memory is unable to furnish.

Some slight corroboration for this explicit, if rather pompously expressed, claim to priority can be found in an article that Sylvester wrote in 1844, referred to in the above remarks as 'my unfinished paper'. This had described various combinatorial systems for which Sylvester had coined the word *syntheme*, the Greek roots indicating a 'grouping together'. His "aggregate systems of triadic synthemes comprising all duads" is what has here been called a complete (3,2) system. Sylvester claimed that the notion of such synthemes had occurred to him when he was an undergraduate at Cambridge in the 1830s.

Whatever the case, this upset Kirkman who became concerned to scotch any suggestion that his presentation of a triple system had derived from Sylvester's work in any way. He had already acknowledged in previous articles that his work had started from Woolhouse's *Diary* question. Soon after Sylvester's historical remarks, he joined in the discussion in a characteristic silky style.

> My distinguished friend Professor Sylvester ... volunteers *en passant* an hypothesis as to the possible origin of this noted puzzle under its existing form. No man can doubt, after reading his words, that he was in possession of the property in question of the number 15 when he was an undergraduate at Cambridge. But the difficulty of tracing the origin of the puzzle, from my own brains to the fountain named at that University, is considerably enhanced by the fact that, when I prepared the question in 1849, I had never had the pleasure of seeing either Cambridge or Professor Sylvester.

Kirkman was not denying – at any rate publically – Sylvester's claim; he seemed to be more anxious to assert and maintain his own independence. For reasons which will be discussed later, he was particularly touchy at this time about the public recognition of his work and there had begun to be a rift between him and the professional mathematicians, like Cayley and Sylvester, with whom he had been until then on friendly terms.

There is no reason to suppose that the schoolgirls had any immediate source other than the investigations that Kirkman made in response to the *Diary* question. What is not clear is whether this question could have arisen from any contacts Woolhouse may had with Sylvester, or indeed one of the latter's 'fellow undergraduates at Cambridge'. It seems to be a curious coincidence that Sylvester's synthemes and Woolhouse's problem first appeared in print in the same year. But Woolhouse would have prepared the material for the 1844 *Diary* in the previous year. This would have been, however, very unlikely to be the source of Sylvester's ideas, for as will be seen later, the notion of a syntheme came from a specific algebraic context that was 'in the air' at the time.

But, with hindsight, there was a further intriguing coincidence in the fact that both men were engaged in actuarial work in London in 1843. Sylvester had just returned from a briefly held university post in Virginia and was taking on occasional actuarial work. Woolhouse had already started work as an actuary with the International Loan Fund. There is no reason to suppose that the men ever met at this time; they came from different backgrounds and led different lives. But Sylvester was a particularly gregarious man, who was always interested in talking about mathematics with a fellow practitioner.

Whatever the case, it is always possible that what was in the air for Sylvester had also been for Woolhouse. So it is worth making a further brief digression. Who was Sylvester? What were the sort of problems he was tackling with his synthemes? What combinatorial problems were of interest to mathematicians at that time?

1.4 Sylvester and synthemes

James Joseph Sylvester was born in 1814 in London, of Jewish parents. The family name, Joseph, was changed to Sylvester when he was young. He had an unhappy childhood, but showed an early talent for mathematics and was an outstanding student at Cambridge. He was ineligible for a higher degree, let alone a fellowship, because at that time graduates were still required to subscribe to the thirty-nine articles of faith of the Church of England.

He did, however, continue with his mathematical work, to start with mainly in applied mathematics, and he was soon appointed to a chair in physics at University College, London. This was one of the few institutions that did not demand theological conformity. Some ten years previously, it had appointed the non-conformist Augustus de Morgan, with whom Sylvester had briefly studied mathematics when he was a boy.

Eventually, in 1841, Sylvester was able to get a post in mathematics at the University of Virginia. This turned out to be an unfortunate move, for hardly had he arrived when he was involved in a quarrel with a student whom he wounded while defending himself with his metal-pointed cane. Various biographers give differing accounts of this incident, which seems to have been hushed up by the authorities once Sylvester tendered his resignation. He seems to have been back in London by 1843, earning a living from actuarial work and private tuition, but still writing mathematical articles.

One of the most important aspects of Sylvester's mathematical career was his association with Arthur Cayley. They met in 1846, when by a strange coincidence they had both decided to study law. At the time, Cayley, who was seven years younger than Sylvester, was also without a job, but was actively engaged in mathematics. Though they were quite different in temperament and in background, they got on well and used to talk mathematics passionately together inbetween their legal studies. Sylvester's interest in mathematics was re-kindled; he was later to refer to his debt to Cayley "for my restoration to the enjoyment of mathematical life".

Both of them completed their legal studies. Cayley was to practice as a solicitor for fourteen years, while producing over two hundred mathematical papers, before he was appointed to the newly established Sadlerian chair at Cambridge in 1863. He married that year and settled down to a productive and uneventful academic career. Sylvester never practised as a lawyer and he never married. He continued to eke out a living with actuarial work and private tuition. After one unsuccessful attempt, he managed to get a post at the Royal Military Academy at Woolwich, where he taught for sixteen years until 1870, when he reached the military retirement age.

Some years later, he accepted the offer of a post at the newly-formed John Hopkins University in Baltimore, where he was to be influential in developing graduate work in mathematics in the States. He was more successful in his teaching than on his previous visit. One of his students later recalled the impact he made in his lectures, capturing something of the excited and excitable style for which he was so well-known.

> The one thing which constantly marked Sylvester's lectures was
> enthusiastic love of the thing he was doing ... We were set aglow by
> the delight and admiration which, with perfect naiveté and with that
> luxuriance of language peculiar to him, Sylvester lavished upon [his
> own] results. That in this enthusiastic admiration, he sometimes
> lacked the sense of proportion cannot be denied. A result
> announced at one lecture and hailed with loud acclaim as a marvel
> of beauty was by no means sure of not being found before the next
> lecture to be quite erroneous ... No young man of generous mind
> could stand before that superb grey head and hear those exposi-
> tions ... without carrying away that which ever after must give to
> the pursuit of truth a new and deeper significance in his mind.

Returning to England when he was 70, he was elected to the Savilian
chair of mathematics at Oxford. But the students there were apparently not
so tolerant of his ways, being more interested in passing their examinations
than in his research work. Eventually he withdrew to London, where he
lived, mainly at the Athaneum club, for a few more melancholy years,
suffering from partial loss of sight and memory. He died in 1897 at the age
of 83.

It is difficult to assess Sylvester's mathematical work. With Cayley and
other British mathematicians, he was a pioneer of a theory of what he called
invariants. He once described these in the following vague, but suggestive,
non-technical terms.

> There are things called Algebraic Forms. [...] They are not, properly
> speaking, Geometrical Forms, although capable, to some extent, of
> being embodied in them, but rather schemes of process, or of
> operations for forming, for calling into existence, as it were,
> Algebraic quantities. To every such [form] is associated an infinite
> number of other forms that may be regarded as engendered from
> and floating, like an atmosphere around it, [...] It is found that they
> admit of being obtained by composition, by mixture, so to say, of a
> certain limited number of fundamental forms [...] the *Covariants* and
> *Invariants* as they are called.

Though invariant theory was an active field of research in the second
half of the nineteenth century, it soon took on an abstract generalised direction where the detailed combinatorial details, with which Sylvester and
Cayley had been involved, were no longer found useful or to be particularly
valued. Sylvester would have understood and sympathised with this generalising tendency, for he always tried to make connections, to place specific

work in a unified setting, to emphasise that "theorems have given place to theories, and no truth is regarded otherwise than a link in an infinite chain".

The last remark came from his Presidential address to the meeting of the British Association for the Advancement of Science in Exeter in 1869. In view of the eventual fate of invariant theory, it is now ironic to read his further observations:

> No mathematician nowadays sets any store on the discovery of
> isolated theorems, except as affording hints of an unsuspected new
> sphere of thought, like meteorites detached from some undiscov-
> ered planetary orb of speculation. The form as well as the matter of
> mathematical science [...] is in a constant state of flux, and the
> position of its centre of gravity is liable to continual change.

Sylvester did not have an easy life, he was always an outsider; but it is not clear how much of this was due to his own temperament and how much to the prejudices of his time. It may be that his mathematical talents were not completely fulfilled. But his often quoted purple passages are valuable reminders of a passion for mathematics that others often conceal.

> Mathematics is not a book confined within a cover and bound
> between brazen clasps, whose contents it needs only patience to
> ransack; it is not a mine, whose treasures may take long to reduce
> into possession, but which fill only a limited number of veins and
> lodes; it is not a soil, whose fertility can be exhausted by the yield of
> successive harvests; it is not a continent or an ocean, whose area can
> be mapped out and its contours defined; [...] it is as incapable of
> being restricted within assigned boundaries or being reduced to
> definitions of permanent validity, as the consciousness, the life,
> which seems to slumber in each monad, in every atom of matter, in
> each leaf and bud and cell, and is ever ready to burst forth into new
> forms of vegetable and animal existence.

The rhetoric disguises, perhaps, some of the complications that can occur in a life devoted to mathematics. It is evident that there were obscure periods in Sylvester's life in which he was perhaps disheartened and disenchanted. It has already been noted that we do not know much about his life immediately after his resignation from the University of Virginia in 1842. In the present context, it would be interesting to know what he was doing in the following two years for this might clarify what might have led him to write an article, in 1844, in which he introduced the notion of a syntheme.

Even if he never met Woolhouse, he might have been influenced by his own actuarial work when he developed this combinatorial notion. With this possibility in mind, there is an intriguing coincidence in the fact that his eldest brother, who had emigrated to the States, was also an actuary. When Sylvester was sixteen, this brother had sent him a problem that was exercising the organisers of lotteries. Sylvester's solution earned him a handsome reward. It is not clear what the problem was, but – to add to the coincidence – Kirkman once mentioned a lottery problem that involved constructing multiples without repeated submultiples, in the manner required by Woolhouse's *Diary* question.

These are but minor themes, however, because although Sylvester's actuarial experience, or imagined contacts with Woolhouse, may not have had any influence on the style of his work at the time, the article in question indicates that he was mainly stimulated by some specific problems in algebra that were being considered at the time. These stemmed from the attempt to find general solutions of algebraic equations, in particular fifth degree ones. A brief account of this issue will be given later. However, the combinatorial notions that Sylvester introduced turned out to be of more general interest and can be considered independently of the particular application he had in mind.

Sylvester's 1844 article was called *Elementary researches in the analysis of combinatorial aggregates* and included one of his characteristic overviews.

> The present theory may be considered as belonging to a part of mathematics which bears to the combinatorial analysis much the same relation as the geometry of position to that of measure, or the theory of number to computational arithmetic; number, place and combination being the three interesting, but distinct, spheres of thought to which all mathematical ideas admit of being referred.

Sylvester had a glimpse of something important, of "theory not theorems"; but he confessed that the difficulty was "to discover proper and principal centres of speculation that may seem to reduce the theory into manageable compass". He hoped to introduce some clarifying order into the contemporary discussion of some problems in algebra through the notion of what he called a *syntheme*. This was defined as "an aggregate of combinations in which all monads appear once and only once". The combinations do not take any account of order, so that *bca* for instance is considered to be the same as *abc*; they may be heterogenous, as in the collection (*abc*, *de*), or – more usually – they may have the same number of elements. Thus "duad synthemes to modulus 4" were sets of distinct pairs formed from 4 elements, such as (*ab*, *cd*), (*ac*, *bd*) or (*ad*, *bc*).

Sylvester also introduced collections of combinations with a uniform repetition of the monads. This notion of a 'balanced' repetition has become important, though it was not taken up by others for some time. He called the set of combinations in which the elements appeared twice and only twice a bisyntheme or diplotheme. Examples might be the heterogenous collection (abc, bc, a), or the quadruple bisyntheme consisting of the letters of the sentence, 'four fast ones hunt here'.

The original syntheme involved duads in which monads occurred once and only once. With the notation in which Kirkman's combinations were previously described, this would be a $(2,1)$ system. To emphasise the frequency of the occurrence of the monads the notation could be extended so that the syntheme is described as a $(2,1,1)$ system, and in this case a bisyntheme ('twice and only twice') would be a $(2,1,2)$ system – there are obvious further extensions, which will be discussed later.

With hindsight, we may wonder why Sylvester did not at this stage also take up the possibility of generalising from monads to pairs, and then triples and so on. It seems such an obvious step to consider "aggregates of combinations" in which all pairs appear once and only once, and so to consider the general combinatorial systems proposed by Woolhouse, and the particular $(3,2,1)$ systems investigated by Kirkman. As noted previously, Woolhouse must have considered this possibility one year before Sylvester's 1844 article. Kirkman was involved with the problem in 1845; it is quite possible that he read Sylvester's paper and thence made the further generalisation himself, but there is no reason to doubt his own claim that he started from the *Diary* question.

It would seem that at this early stage in the study of general combinatorial structures, it was not necessarily very clear to the pioneers which of the notions being entertained were going to be the most important. In some ways the professional mathematician – for so Sylvester would have considered himself, despite his lack of a university post at the time – was hampered by his knowledge of current issues. Thus, he would have been looking for notions that would clarify existing problems such as the solution of fifth-degree equations, and, perhaps, less likely to see a future potential. On the other hand, the amateur – and this is how Kirkman would have seen himself at the time, despite his later ambitions – would be perhaps freer to explore structures for their own sake, and in doing so was sometimes able to open up more creative possibilities for further work.

It is a poignant comment on Sylvester's passionate interest in the continuing re-organisation of mathematical knowledge that he seems to have missed the real significance of his own synthemes. Many years later, in an article containing some historical remarks about the schoolgirls problem that upset Kirkman, he slipped in the further development that he had not made in 1844.

A syntheme, I need hardly add, is an aggregate of combinations containing between them all the monadic elements of a given system each appearing once only. In the more general theory of aggregation, such an aggregation would be distinguished by the name of a monosyntheme. A disyntheme would then signify an aggregate of combinations containing between them the duadic elements, each appearing once only, and so forth.

Thus the systems of multiples without repeated pairs that he had not discussed in 1844 were now called disynthemes – in the notation suggested above, these were $(y,2,1)$ systems, whereas his bisynthemes were $(y,1,2)$ systems. Sylvester was now able to refer to the schoolgirls problem as requiring the construction of "a triadic disyntheme, separable into monosynthemes to the base 15" – almost as if this is what he had had in mind at the time! In switching from the Latin prefix 'bi-' to the Greek prefix 'di-', he may have betrayed that he had not really seen the significance of his own notion of uniform repetition of monads. At the same time his claim to some historical priority in the conception of systems without repeated pairs becomes a little doubtful.

It is a form of hindsight to be looking at Sylvester's work in terms of its eventual significance for later combinatorial studies, or its possible links with Kirkman's investigations into triple systems. In fact, Sylvester can be more closely linked to Kirkman through their independent contributions to the emergence and development of the mathematical notion of a group. This development may be said to have started mainly through various mathematicians' work on the general solution of algebraic equations. A brief account of this background, which may help to place Sylvester's 1844 article in its proper context, is given in a later section 4.2.

Further combinatorial investigations

2

XXX. *Theorems in the Doctrine of Combinations.* By the Rev. Thomas P. Kirkman, *A.M., Rector of Croft with Southworth.*

To the Editors of the Philosophical Magazine and Journal.

GENTLEMEN,

I WOULD beg your permission to enunciate the following theorems in your Journal :—

A. With 7 symbols can be formed 21 triads, so that every duad shall be thrice employed.

B. Two distinct systems of 7 quadruplets each can be made with 7 symbols, both exhibiting twice all the 21 duads.

C. A system of 21 quadruplets can be made with 7 symbols, so that every possible duad shall be six times employed.

D. With 13 symbols can be made three different groups of triads, each group once containing all the duads.

E. With 15 symbols different triads can be made, so as to exhaust the possible duads, once, twice, 3, 4, 5, 6, 7, 8, 9, 10, 11, 12, or 13 times.

F. With $(12n+3)$ symbols can be formed triads so as to exhaust the duads $6n-1$, or $6n+2$ times; and with $12n+7$ symbols, so as to exhaust the duads $6n+1$, or $6n+3$ times.

G. With 27 symbols triads can be made, till the duads have been all twice employed, or all thrice employed.

H. With $4(3n+1)$ symbols quadruplets can be made, till every duad has been $(2n+1)$ times employed, and this without repeating any triplet.

I. With 4×2^n symbols, quadruplets can be made till every triad has been once employed.

J. Sixteen young ladies can all walk out four abreast, till every three have *once* walked abreast; so can thirty-two, and so can sixty-four young ladies; so can 4^n young ladies.

Croft Rectory, near Warrington,
August 6, 1852.

Phil. Mag. S. 4. Vol. 4. No. 24. Sept. 1852.   P

2.1 Quaternions, pluquaternions and sums of squares

When Kirkman had completed his first mathematical publication, it was natural for him to look for other problems he might tackle. In the isolation of his vicarage study, this would mean browsing in the current journals. So it is not surprising to find that he was drawn to the study of quaternions, which were being assiduously investigated by many mathematicians since these new 'numbers' had been first created by Hamilton in 1844. More than 70 articles on the subject were published in British journals during the 1840s. Many of these articles were by Hamilton himself, whom Kirkman would have known from his student days in Dublin.

William Rowan Hamilton was acknowledged to be the Newton of his age, or as one contemporary called him, the Irish Lagrange. Born in Dublin in 1805, he showed an early talent for mathematics, and was appointed in 1831 to the chair of astronomy at Trinity College, while still technically an undergraduate. He made some important contributions in optics, dynamics, and the calculus of variations; he received various academic honours and was knighted in 1835. The second part of his life was devoted to quaternions, against a background of a struggle with alcoholism and an increasingly unhappy personal life. He died of gout in 1865 at the age of 60.

What were quaternions and why should there have been so much contemporary interest in them? Principally, they were an extension of the notion of complex numbers, which were themselves extensions of the ordinary numbers. Complex numbers had appeared in the form of 'fictitious' solutions of certain quadratic equations; for example, $x^2 + 1 = 0$ which was at first solved merely formally by writing $x = \sqrt{(-1)}$. By the 1830s, various ways of interpreting these imaginaries had been proposed. Whereas ordinary numbers could be represented as magnitudes on a line – a so-called number-line – the complex numbers were now represented by direct magnitudes in a plane. Hamilton had also proposed a further interpretation of a complex number as an operator on vectors in a plane. Thus the square root of minus one could be interpreted as a rotation through a right angle, for when this is applied twice to a positive unit vector the result is a negative unit vector.

Some mathematicians were not very happy about basing complex numbers on a geometrical interpretation, however intuitively satisfying this might be. The

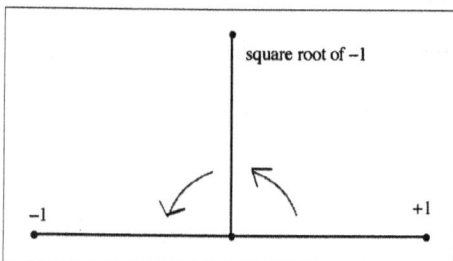

discovery of non-Euclidean geometries at the beginning of the nineteenth century meant that people started trying to separate ideas about numbers from their natural geometrical connections. In 1837, Hamilton proposed a purely algebraic representation of complex numbers as ordered number-pairs, namely pairs of numbers (a,b) which were capable of being combined in certain defined ways, called addition and multiplication. The formal definitions of these operations drew attention to the associative, commutative and distributive laws, which had until that time always been unconsciously assumed. Mathematicians began to be more aware of these algebraic laws as they began to develop new number-systems.

In what sense was the square root of minus one a number? Was it justifiable to call a pair of numbers a number? What is a number for that matter? According to Oswald Spengler, "there is no such thing as number as such – only several number worlds". Number are made by humans and notions about them change. Fractions were studied by the Greeks as ratios of commensurable magnitudes, and what are these ratios but number-pairs? But these could still be represented on a number-line. Until the nineteenth century, numbers remained one-dimensional, they could be ordered and represented geometrically by points on a line. The square root of minus one could not: it had to be represented on a plane. Complex numbers are essentially 2-dimensional.

The question naturally arises whether there are numbers that are 3-dimensional. Are there number triples representing points in 3-dimensional space which in some sense or other could be justifiably called numbers?

This problem seized the imagination of Hamilton and he spent many years trying to find a solution. The trouble was that though it was easy enough to add triplets, it is not so easy to combine them in any way that is recognisably a multiplication. It is said that his young children used to ask him at breakfast if he could yet multiply triplets, whereupon he would sadly admit that he could still only add and subtract them. A breakthrough came suddenly one day in 1843, when Hamilton realised that whereas an operation on a vector in the plane could be specified by a number-pair, a vector operation in ordinary space required a number-quadruplet.

He associated the four numbers with four units, namely the number 1 and three 'imaginaries' i, j, k, each of which was a square root of minus one in the sense that their squares were equal to one. He called an expression of the form $a+b.i+c.j+d.k$ a quaternion. This word had been used before – in algebra to mean a product of four terms, in printing for a sheet of paper folded twice, and in the English Bible to describe a group of four soldiers.

Hamilton still found that he could not combine these in a way that retained all the characteristics of multiplication. But he did find a rule that retained all but the commutative property, and his second breakthrough was to adopt this boldly as a non-commutative multiplication. In terms of the units this meant that expressions like $i.j$ and $j.i$ were not equal as in ordinary multiplication, but that one was the negative of the other. Such relationships are now common, not only in pure mathematics but also in physics. But they were very startling at the time they were first introduced. Hamilton once described a particular interpretation:

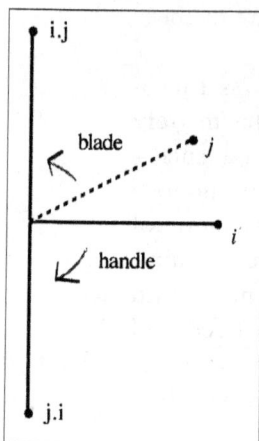

> By taking from my pocket a penknife and partly opening it in a horizontal posture, whereof for the moment we may agree to call the handle i and the blade j, I showed that by operating on j with i, by turning the blade through a quadrant with a screwing action, that blade was made to point upward. Whereas on the contrary when I operated on i with j, or used the blade as the axis of the screwing motion, the handle was made to point downwards.

A useful modern equivalent of this image could be the Rubik Cube in which the non-commutative nature of the rotations are immediately noticed through the different colour arrangements that arise. This image also indicates one of the difficulties that people had when trying to interpret quaternions. If the 'imaginaries' were rotations in space, what was the real number unit? Hamilton did occasionally speculate that this was a sort of fourth dimension of time. But most of his contemporaries were rather suspicious of his metaphysical interest in time – he had previously published an account of algebra as a science of time – and this early hint of the space-time of modern relativity theory was not pursued.

In a famous account of his discovery, written more than twenty years later in a letter to his son, Hamilton recalled walking along a canal with his wife on a particular day:

> An undercurrent of thought was going on in my mind, which gave at last a result, whereof it is not too much to say that I felt at once the importance. An electric circuit seemed to close; and a spark flashed forth, the herald (as I foresaw, immediately) of many long years to come of definitely directed thought and work… Nor could I resist the impulse – unphilosophical as it may have been – to cut with a knife on a stone of Brougham Bridge, as we passed it, the fundamental formula with the symbols i, j, k; namely $i^2 = j^2 = k^2 = ijk = -1$, which contains the solution to the problem.

Hamilton lectured on his discovery at a meeting of the Royal Irish Academy a month after his walk on the canal bank. He published an article in the following year 1844, and continued to publish many more for the rest of his life, which he devoted almost exclusively to the study of quaternions. As it happened, these were superceded by other more general developments, in the same sort of way that Sylvester's invariants were. But quaternions had an immediate impact, rather like that of the non-Euclidean geometries that were being discovered at about the same time. Many people were disturbed by the displacement of a traditional understanding of multiplication. Even John Graves, who was a sympathetic colleague and former fellow-student, had some reservations at the beginning:

> There is still something about the system that gravels me. I have not
> yet any clear views as to the extent to which we are at liberty
> arbitrarily to create imaginaries, and endow them with supernatural
> properties. You are certainly justified by the event. You have got an
> instrument that facilitates the working of trigonometric theorems
> and suggests new ones and it seems hard to ask more; but I am glad
> that you have glimpses of physical analogies.

But Hamilton's previous achievements had earned him an international reputation and many of his contemporaries, including Graves, began to study new number-systems with enthusiasm. Quaternions specified operations on vectors in ordinary space, but were represented by quadruples, so that they could be thought of as 4-dimensional numbers. These had been created at the expense of a fundamental law of multiplication. It seemed likely that further extensions would require further sacrifices and, indeed, Graves and others found one in the form of a system of 'octuples' that required a non-associative multiplication (in this case, the product of three terms is ambiguous – the order in which the multiplications have to be performed is not irrelevant and has to be specified). These octaves or 'octonions' could be thought of as 8-dimensional numbers; they are often named after Cayley who discovered them independently in 1845.

One interesting feature of quaternions that people wanted to retain when seeking extensions to further dimensions was a simple consequence of the non-commutative multiplication; indeed, this multiplication had been chosen partly in order that quaternions should have this feature. The property in question was itself an extension of a similar one for complex numbers.

When a quaternion $a+bi+cj+dk$ is multiplied by its so-called conjugate, $a-bi-cj-dk$, the result is a number without any imaginary terms. This may be shown by formal manipulation of the rules for multiplying imaginaries.

$$(a+bi+cj+dk).(a-bi-cj-dk) = (a^2+b^2+c^2+d^2) - (ab-ba)i-bc(ij+ji) - \dots \text{ etc}$$
$$= (a^2+b^2+c^2+d^2)$$

Thus, the product of a quaternion and its conjugate is a sum of four squares. Conversely, such a sum may be factorised into the product of a quaternion and its conjugate. This means that two sums of four squares have a product that can be factorised into four factors which can then be recombined in an intriguing and satisfying way.

$$(a^2+b^2+c^2+d^2)(A^2+B^2+C^2+D^2)$$
$$=(a+bi+cj+dk)(a-bi-cj-dk)(A+Bi+Cj+Dk)(A-Bi-Cj-Dk)$$
$$=(a+bi+cj+dk)(A+Bi+Cj+Dk)(a-bi-cj-dk)(A-Bi-Cj-Dk)$$
$$=(W+Xi+Yj+Zk)(W-Xi-Yj-Zk)$$
$$=(W^2+X^2+Y^2+Z^2) \text{ with } W=aA-bB-cC-dD, \text{ etc}$$

Thus the product of two sums of four squares can be expressed as a sum of four squares. This four-square formula was first discovered by Euler in his proof of a celebrated conjecture by Fermat that every integer could be expressed as a sum of four integral squares. It is a generalisation of the classical similar formula for two squares.

The four-square formula had already been generalised to eight squares before quaternions were invented. It was quickly re-discovered by Graves and Cayley, who went on to investigate extensions to 16 squares. This would be possible if there were a system of 15 imaginaries which are all square roots of minus one and which are such the product of any pair is a single imaginary. The problem is then to find a system of mutually consistent products. Cayley noticed that since any pair of imaginaries had to be associated with some unique imaginary, then the imaginaries had to form a system of triples with repeated pairs! He recognised that this would be of interest to Kirkman and wrote to him about it sometime in 1848. Kirkman was stimulated to write an article, *On pluquaternions and homoid products of sums of squares*, by the end of that year.

> The following analysis is the fruit of my meditations on Professor Sir W. R. Hamilton's elegant theory of quaternions, and on a pregnant hint kindly communicated to me without proof by Arthur Cayley, Esq., Fellow of Trinity College, Cambridge, about the connexion between a system of triples having no duad in common and the property that the product of two sums of $2n$ squares shall be a sum of $2n$ squares.

Kirkman began by considering closed systems of imaginaries $i, j, k \dots$, which were all square roots of minus one. In such cases, multiplication is not commutative and the product of any two imaginaries in one of its orders is some other unique imaginary, say $i.j=k$. This means that the imaginaries form a system of triples i, j, k without repeated pairs. Kirkman could then deduce from his previous work that the number of such imaginaries must be of the form $6k+1$ or $6k+3$. Complex numbers had one imaginary, quaternions had three. Kirkman called any further extensions *pluquaternions*, "not dreading here the pluperfect criticisms of grammarians, since the convenient barbarism is their own".

In constructing a system with seven imaginaries – like Grave's octaves – Kirkman found that multiplication with the particular non-commutative rule $i.j=-j.i$ could not in general also be associative, for otherwise $i(jk) = (ij)k = -k(ij)= (-ki)j = (ik)j = i(kj) = -i(jk)$, in which case $i(jk) = 0$ and every triple product would have to be zero. He constructed various possible systems with seven imaginaries. These all involved a non-associative rule for any triple product by which $i(jk) = -i(jk)$. He then showed that to extend the original system of seven imaginaries required at least a further three, but that in this case the system involved a contradiction.

One of the possible systems with seven imaginaries may be presented in terms of the defining relations: $i = jk = wx = zy, j = ki = wy = xz, k = ij = wz = yx$. If two or more imaginaries, say q, r, are included in the system, then the product of these must be some single previous element, say $qr=i$. This means that r is linked with i and q. Since j is already linked with the original elements, it follows that the pair r, j cannot be linked with any element other than a further one, say $jr=p$, whence $j=rp$ and $k=pq$. Hence to extend the original system of seven imaginaries requires at least a further three, p,q,r. But since $i = zy = qr \rightarrow (zy)r=(qr)r \rightarrow z(yr)=q(rr) \rightarrow (zz)(yr) = zq \rightarrow (1): yr = -zq$, and similarly $j = wy = rp \rightarrow (2): wp = -yr$ and $k = wz = pq \rightarrow (3): wp = -zq$, and the three numbered conclusions are contradictory.

It was thus impossible to have a closed system of pluquaternions with more than seven imaginaries. The product of two pluquaternions could not then be a pluquaternion. It followed that in general the product of two sums of n squares could only be a sum of n squares for $n=2,4,8$.

But Kirkman was interested in various other ways of extending the classical four-square formula. He went on to discuss the ways in which the product of two pluquaternions could be a pluquaternion of a different dimension. In a continuation of his article, which was published later in the same year, he deduced, with formidable detail, various theorems about products of sums of an equal number of squares. To emphasise the similar form of the two sums, he called these *homoid* products in the eye-catching title of his article. Certain homoid products were equal to sums of other numbers of squares. For example, in general the product of sums of 12 squares is a sum of 26 squares, and the product of sums of 16 squares is a sum of 32 squares. Some of his results were based on conjectures about which he was a little uneasy and he concluded the article with a plea for the reader's indulgence on the curious grounds that anyone other than Hamilton would find it very difficult to be completely clear about the issues involved.

> For whatever is obscure, unfinished, or even illogical, I trust to
> receive every indulgent allowance to which the confessed difficulty
> of these subjects may entitle those, who, while they are not
> forbidden to speak on them, are yet not expected to bring to their
> discussion the powers of mind possessed by such writers as the
> distinguished inventor of quaternions, or to exhibit his brilliant
> results in rich and varied applications.

Kirkman's theorems on homoid products have been more or less ignored; once again, the particular theorems of a pioneer have been by-passed by the development of more general theories. His other main result, that there were no further closed systems of pluquaternions after Cayley's octonions, was also to be approached in a different way. Cayley had already realised that the dimensions of any further extension would have to be a power of 2; and he had suspected that there were no extensions to 16 dimensions. But Cayley was unable to show this by proving that his condition failed for a system of 15 imaginaries. Indeed, it seems likely that he wrote to Kirkman specifically to enlist his aid.

As Cayley had realised, a system of 15 imaginaries would have to be a system of triples without repeated pairs. Kirkman was bound to draw on his original paper, especially his actual construction of a (3,2) system on 15 elements. This particular emphasis on fifteen must surely be part of the background to his selection of that number for the schoolgirls problem that he invented the following year. He certainly continued to be particularly interested in the triple system for 15 elements; and it re-appeared in all the five articles of his that were published in 1850.

In one of these articles, he introduced a curious notation to deal with the rather complicated book-keeping that could arise from the particular rules for the multiplication of imaginaries. The product of two imaginaries i, j will be equal in value to a third, say k, but with a positive or negative sign depending on the order in which the multiplication is taken; a similar ambiguity will arise in taking triple products. Kirkman usually dealt with such ambiguities by using the standard plus-or-minus notation. But the usually convention is that a choice of a particular sign extends throughout any particular expression. In the case of triple products of imaginaries the ambiguous signs were sometimes independent. So Kirkman placed such ambiguities in brackets. He called a quantity "whose value was independent of its sign, so that it may be employed in the same argument with contrary signs" a *bisignal univalent*.

With this notational device, Kirkman was able to offer a brief account of the product of two sums of 16 squares and the constraints that there would have to be on the numbers involved in order for this product to be itself a sum of 16 squares; he added a brief similar discussion for 23 and 31 squares. It was acknowledged that the result for 16 squares had been previously stated by John Young, but "the author gives no account of the process by which he arrived at the result". Kirkman must have felt a little uneasy about his own treatment; he appeared to argue, quite unmathematically, that the ends justified his means.

> Should the reader hesitate to accept this conclusion on the strength
> of my arguments, from a very natural suspicion that the latter are
> more to be admired for their luck than their learning, I trust that
> he will allow the result to be an apology for the reasoning, and be
> lenient to the logic for the conclusion's sake.

He went on to express the hope that his bisignal imaginaries might turn out to be more than "fortunate nonsense and two useless words" when the theory of imaginaries was developed into "an algebra of time and order". Certainly, theorems about products of sums of squares are now seen in the more general context of algebras that may have a non-associative multiplication or one that has no corresponding division. By the end of the nineteenth century, it had been shown that the product of n squares would only be in general a sum of n squares when $n=2$, 4 or 8, and that these were the only possible dimensions for any division algebra, though it is doubtful whether many modern mathematicians would see these as algebras of time and order.

2.2 Multiple systems, projective planes and latin squares

In his 1846 article, *On a problem in combinations*, Kirkman had completely solved the problem of determining necessary and sufficient conditions that a number of elements can be arranged in triples that cover all pairs in the sense that every pair of elements occurs in one and only one pair. Such an arrangement can be called a triple system and Kirkman had shown that triple systems exist if and only if the number of elements is of the form $6k+1$ or $6k+3$.

It was natural for Kirkman to turn his attention to quadruple and further multiple systems. In an article published in 1850, *Note on an unanswered prize question*, he observed that there had been no response to the general problem posed in the *Lady's and Gentleman's Diary* for 1844, other than his own discussion of triple systems. He then presented some new theorems on multiple systems, namely $(y,2)$ systems of sets of y elements covering all pairs, for any y. (He actually used r and $r+1$ rather than the y used here.) The first of these theorems asserted that there are arrangements of y-sets (that is, sets of y elements) which cover, in the sense described above, all pairs when y is prime and the number of elements is a power of this prime y. Thus, for example, there are triple systems with 3, 9, 27 ... elements (this would also follow from his original results since these numbers are all of the form $6k+3$), and there are quintuple systems with 5, 25, 125 ... elements.

A second theorem asserted that there is a system of y-sets that cover all pairs of y^2-y+1 elements, for prime $(y-1)$. Thus, there are quadruple systems with 13 elements, sextuple systems with 31 elements and so on. This theorem presented an interesting combinatorial feature that Kirkman did not comment on. Writing the number of ways of choosing pairs from n elements as $C(n,2) = n(n-1)/2$, the number of y-sets that cover all pairs of y^2-y+1 elements is $C(y^2-y+1,2)/C(y,2)$, which is precisely y^2-y+1 as Kirkman showed. But this creates a duality in the system; by its very definition each pair of elements appears in one and only one y-set, but in this case every pair of y-sets have one and only one element in common.

This duality is a feature of projective geometry and it can now be seen with hindsight that Kirkman had created a configuration of elements that was later to be called a *finite projective plane*. In this case, the elements are interpreted as points and the y-sets as lines, though these are not the points and lines of ordinary euclidean geometry. The duality, if it exist, means that every line contains y points and every point lies on y lines. Taking one of these points there are then y lines passing through it each of which contains $(y-1)$ other points. The total number of points is therefore $1+y(y-1)$, namely y^2-y+1.

In a projective plane any two lines always intersect (the y-sets have one and only one element in common) so that there are no parallels. When one of the lines is singled out specially (for example as the 'line at infinity') then two other lines that intersect on this special line are called parallel and the plane is called an affine plane. Now consider a finite projective plane with y points on each line and delete one line with its y points. Each of the $(y-1)$ other lines originally met the deleted line in one, and only one, of the deleted points, so that these other lines now each only have $(y-1)$ points on them. They therefore form a system of $(y-1)$-sets covering all pairs of $(y-1)^2$ elements, now called a *finite affine plane*. Clearly such a plane exists if and only if there is a corresponding finite projective plane with y points on each line.

Finite projective and affine planes are all of the form described, but to determine the values of y for which they actually exist is a difficult and unresolved problem which has been extensively studied in recent times. Kirkman's first theorem implied that there are finite affine planes with $(y-1)$ points on each line whenever $(y-1)$ is prime. He deduced his other theorem by precisely the equivalent of restoring the 'deleted line' to assert – in a different language – that there are finite projective planes with y points on each line whenever $(y-1)$ is a prime.

These results were neglected by contemporary mathematicians, perhaps as Kirkman himself suspected, because his work was affected by his reputation as a poser of trifling problems about schoolgirls. Thus, despite the contributions of eminent mathematicians like Cayley and Sylvester to the original schoolgirl problem, the author of an article on the topic in the *Proceedings of the London Mathematical Society* for 1880 apologised to members of the society for troubling them with "nothing but an ingenious puzzle".

Kirkman's theorem on y-sets with $(y-1)$ prime was first stated in terms of finite projective planes when it was re-discovered by Guiseppe Fano in 1892. The simplest case, when $y=3$, is now known as the Fano plane: the figure below shows its traditional representation as a set of 7 points and 7 lines, with the line XYZ indicated by a circle. The other figure shows the corresponding affine plane as a tetrahedron, where XYZ is taken to be the line at infinity and the parallels are skew lines, like AD and BC.

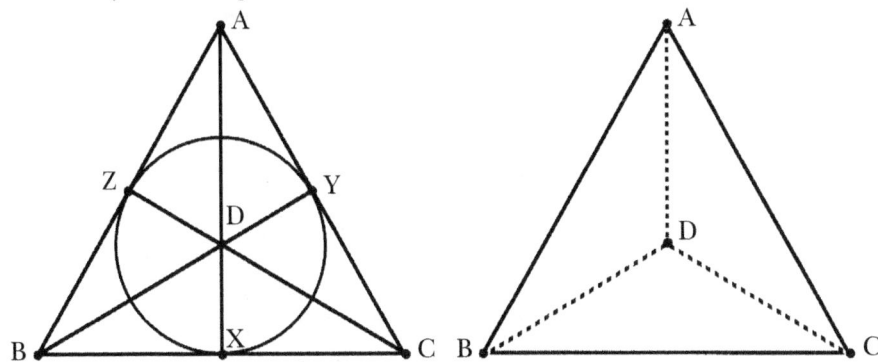

In fact, it turned out that finite projective planes exist whenever $(y-1)$ is a power of a prime. It follows that there are finite projective planes with y points on each line for $y = 3,4,5,6,8,9,10,12$ and so on. In the other direction, it has been shown that there are no planes when y is of the form $4k+2$

or $4k+3$. This complicated condition eliminates an infinite number of values of y, for example 7,15,22,23. But these results still leave an infinite number of unresolved cases. The first of these is when $y=11$ and it remains an outstanding problem to prove or disprove the existence of a finite projective plane with 11 points on each line.

The elimination of the case when $y=7$ was implicit in the work of a French mathematician, Gaston Tarry, who in 1900 solved a famous problem about 36 officers posed by Euler in 1782. This problem required the arrangement of 36 officers of 6 different ranks and from 6 different regiments in a square such that each row and each column contains one and only one officer of each rank and one and only one officer from each regiment.

In any solution of the problem the array of ranks alone, or similarly the array of regiments alone, would be an example of what Euler called a *Latin square*. This is an array of elements, in this case 6 of each of 6 different types, which appear once and only once in each row and column (other examples include the recently popular Sudoko puzzles like the completed one shown below). Two Latin squares that can be paired in the manner required by the problem about officers are said to be *orthogonal*. Euler had conjectured that there are no orthogonal pairs of n-by-n Latin squares when n is of the form $4k+2$, thus asserting that the problem of the 36 officers was insoluble.

4	8	3	9	5	7	6	2	1
2	1	9	6	3	4	7	5	8
5	7	6	1	8	2	9	4	3
8	2	5	7	4	6	3	1	9
7	6	4	3	9	1	2	8	5
3	9	1	8	2	5	4	7	6
1	5	7	4	6	9	8	3	2
9	4	8	2	1	3	5	6	7
6	3	2	5	7	8	1	9	4

Tarry verified the conjecture for $n=6$. For some years this result and the investigation of Latin squares in general remained a topic of recreational mathematics. But in the nineteen-thirties statisticians began using Latin squares in the design of biological experiments. If the growths of some varieties of seed, for example, are to be compared, then it becomes important to spread each variety evenly over the plot so that special features of say soil or drainage do not favour one type over another. Latin squares provide one of the simplest ways of ensuring an even spread and they began to be studied more seriously as a result. To determine the effect of a number of fertilisers on the varieties of seed requires an even interaction of fertiliser and variety and the appropriate design would be a pair of orthogonal Latin

squares. The picture on the right shows a 5×5 Latin square of trees planted in a forest in Wales in 1929 to compare the effect of exposure of five different types of tree.

Then in 1938 an Indian statistician, Raj Chandra Bose, related the existence of projective planes with y points on a line to the construction of orthogonal $(y-1)$-by-$(y-1)$ Latin squares, so that Tarry's result now showed that there were no finite projective planes with 7 points on each line.

Bose later collaborated with others to prove that there are orthogonal n-by-n Latin squares for every n of the form $4k+10$, thus turning Euler's conjecture in the opposite direction and providing a neat example of how the greatest of mathematicians can draw the wrong conclusions from limited inductive evidence.

Meanwhile, surprisingly little further advance has been made in the general study of Kirkman's systems of multiples covering all pairs, especially in establishing sufficient conditions on the number x of elements in a system. Necessary conditions are easy to find. Thus, if a particular element A occurs in Y of the y-sets, then there are $x-1=Y.(y-1)$ pairs of elements containing A. This means that Y is invariant, that is that *all* elements occur in Y of the y-sets and it is necessary that $Y=(x-1)/(y-1)$ be integral. Furthermore, if there are X of the y-sets in the system, then counting the number of elements in two ways gives $X.y=Y.x$ and it is also necessary that $X=x(x-1)/y(y-1)$ be integral. Similar necessary integral conditions occur for the more generalised combinatorial structures that were later developed from Kirkman's multiple systems.

The conditions mean that a $(y,2)$ system must have a number x of elements such that $(x-1)/(y-1)$ and $x(x-1)/y(y-1)$ are both integral. Explicit conditions for x may be derived for particular y by testing all remainders of x when divided by $y(y-1)$. For example, when $y=3$ the conditions are that

$(x-1)/2$ and $x(x-1)/6$ are both integral: the remainders of x when divided by 6 are called the residues of x mod 6 (where mod is an abbreviation for 'modulo', meaning 'measured by'); and in this case the only possible residues are 1 or 3, so that in case the conditions are only true when x is of the form $6k+1$ or $6k+3$. It will be recalled that Kirkman had shown in 1847 that these were also sufficient.

When y is any power of a prime, it can be similarly shown that $x=1$ or y mod $y(y-1)$. Kirkman constructed various systems with particular elements of this form in 1850. More than a hundred years later it was shown that there are always systems with any number of elements of this form when $y=4$ or 5. When y is composite, the situation is more complicated. Thus, when $y=6$ the testing of residues mod 30 gives four possibilities for x, namely 1,6,16 or 21. These conditions cannot be sufficient as they stand, for there are no systems formed with some of these values. Kirkman's results for prime $(y-1)$ establish the existence of a sextuple system with 31 ($=1$ mod 30) and 186 ($=6$ mod 30) elements. It is also known that there are systems with 46 and 51 elements, so that the two extra necessary conditions are certainly relevant. But it is known that there are no sextuple systems covering pairs of 16, 21 or 36 elements.

A system of the last sort was long sought in connection with a generalised schoolgirl problem posed by Benjamin Peirce involving the scheduling of walks for y^2 girls in groups of y. Rouse Ball's book on mathematical recreations gives solutions for various values of y, but reports that the case when $y=6$ had "baffled all attempts to find a solution". But a system of sextuples with 36 elements would be a finite affine plane correspondng to a finite projective plane with 7 points on a line and this is excluded by Tarry's result.

The difficulty – now partially resolved – in establishing conditions that would be sufficient to guarantee the existence of systems of multiple covering pairs does not detract from their importance or interest. Their essential feature, which was to be be taken up later by statisticians, is an even 'balanced' distribution of pairs over the elements which, as the counting argument shows, induces a balanced repeated distribution of the elements themselves. This balance in the structure could be specified by the number X of actual multiples, the number Y of those that contain any one element, and the number – in this case 1 – of those that contain any one pair of elements.

As has been already shown, the numbers X and Y are invariant and dependant on the number x of elements, and the number y of elements in each multiple. More general balanced systems can be constructed by having more than one multiple containing any one pair of elements.

2.3 Cubic, quartics and Steiner triples

The development of algebraic tools for the study of geometry – principally
by German mathematicians in the first half of the nineteenth century – led
to an analysis of cubic, quartic and higher-degree curves. Many mathematicians became intrigued by the configurations formed by certain special
points on such curves.

Julius Plücker in his book, *System der Analytische Geometrie* (1834), showed
that a cubic curve had at most three real inflexions (namely, points where
the tangents begin to swing the other way, as in the figure below), but that
when imaginary points were taken into account there were always nine
inflexions and these lay in threes on twelve lines. This provides an example
of a triple system with nine elements. Sylvester referred to this system as the
source of his interest in certain alignment problems (discussed in a later
section, 3.4), now known in books on mathematical recreations as tree-
planting or orchard problems.

A few years later, Plücker also showed that a quartic curve had at most
eight real bitangents (namely, double tangents touching the curve twice, as
in the figure below), but that when imaginary lines were taken into account
there were always 28 bitangents and these lay in fours whose eight points of
contact lay on a conic.

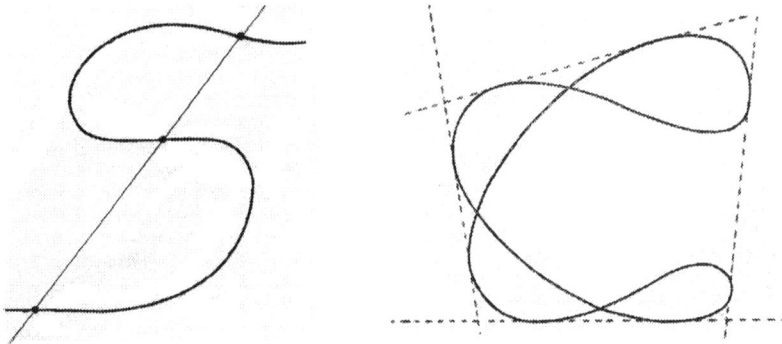

Meanwhile there were other mathematicians working independently in
the same areas of interest. It seems at first sight to be yet another example
of independent and simultaneous discovery in science that at about this
time Plücker's great rival, Jakob Steiner, who had initiated the study of the
Pascal configuration and had many other important contributions to
geometry, also became involved with the combinatorial systems investigated
by Kirkman. In 1853, he published a brief note, *Combinatorische Aufgabe*, in
a German journal which was always called Crelle's Journal after its founder
and first editor, August Crelle. The paper began with the question:

What number of elements has the property that the elements can be grouped in threes so that any two elements always appears in one and only one combination?

This asks in effect for a covering by triples of *all* pairs and Steiner asserted – correctly, as Kirkman had already shown – that this is only possible when the number of elements is of the form $6k+1$ or $6k+3$. Steiner went on to pose some generalisations involving quadruples and other multiples and ended with some remarks to explain how he came to be involved with such problems. In this case, there is an explicit link with Plucker's configurations.

I chanced upon this problem about six years ago, when studying the double tangents of curves of the fourth degree. This geometrical observation cast light on the nature of the required combinations but was not enough to explain them fully.

This means that Steiner would have been working on the problem of enumerating multiples that cover all pairs at about the same time as Kirkman, whose first paper on the topic had been published in 1847. This would be a neat coincidence, not unusual in mathematics, that could easily be explained by the fact that geometrical configurations were of interest to many mathematicians at that time and that some common combinatorial characteristics would be bound to be identified sooner or later.

There are, however, some curious features of the situation in this case, Steiner never published any other work on the topic. Moreover, he was fifty-six years old when he published his note and it was known that in his later years he was worried about his waning creative powers and anxious about his reputation. It has been claimed that he took over theorems and proofs published in English and other journals without mentioning that his results had already been established.

Steiner was an unusual man. The son of a Swiss farmer, he was largely self-educated and did not learn to write until he was fourteen. In his early years he taught in the school at Neuchâtel run by the famous educational reformer, Johann Pestalozzi, and he retained a lively interest in pedagogy when he became a professor in Berlin. He was passionately devoted to synthetic (ie non-algebraic) methods and to the development of geometrical intuition, so much so that he would teach geometry without drawing figures and in a darkened room.

He was so antipathetic to algebraic methods in geometry that he once threatened to stop writing for Crelle's Journal if Plücker's work continued to appear there. His immense contemporary reputation ensured that his critical reception of Plücker's work caused it to be neglected. Plücker, who with hindsight was the more profound geometer, became very discouraged and for a time retired from mathematics and devoted himself to physics. He did return to mathematics in his later years, largely due to the response given to his work by English mathematicians.

If Steiner did plagiarise in this case – and this is unproven – this seems a relatively harmless matter only of concern to pedantic historians. Though it clearly irritated Kirkman, who wrote: "How did the *Cambridge and Dublin Mathematical Journal* contrive to steal so much from a later paper in Crelle's Journal on exactly the same problem in combinatorics?" Whatever the case, tradition has firmly ignored Kirkman's clear priority and system of triples covering all pairs are now known as *Steiner triple systems*.

The traditional account is more deeply mistaken in that it emphasises that Steiner asserted a necessary condition and that it was M. Reiss who showed in 1859 that it was also sufficient, that is that triple systems exist for all numbers of the form $6k+1$ or $6k+3$. But this too had been established by Kirkman in his combinatorial paper of 1847. His method was similar to that given by Reiss and it has been suggested that the latter may also have borrowed from Kirkman's paper.

Triple systems have been studied extensively. The simplest is a set of three elements; the next largest is a set of 7 triples of 7 elements, a unique system that can be represented in various ways – for example,

- by the following three-letter words: ADO, BAR, BED, BOY, DRY, ORE, YEA;
- by three-digit binary numbers (other than zero) arranged in groups of three whose corresponding places add to zero (for example, 011, 101, 110);
- by interpreting the latter groups of three as winning positions in the game of Nim with three piles of up to seven counters in each pile (for example the previous triple would be a winning position 3,5,6);
- the integers mod 7, grouped into the quadratic residues (1,2,4) and six more triples obtained by successively adding one to each term – so (2,3,5), (3,4,6) and so on.

The next system is also unique – a set of 12 triples of 9 elements which can be represented by complex inflexions of a cubic curve. Larger systems are not unique; there are, for instance, two distinct systems for 13 elements and eighty systems for 15 elements. The number of distinct systems increases very rapidly for larger number of elements.

Meanwhile triple systems have also been considered from another algebraic point of view. Since there is a unique triple containing any pair of distinct elements, a binary operation is defined on the elements of the system. This operation is not in general associative, but any two elements of a triple uniquely determines the third so that the elements form what is known as a quasi-group.

2.4 Perfect partitions, cyclic projective planes and difference sets

Some members of the Lancashire and Cheshire Historical Society may have been a little startled on May 21st 1857 when the Rev. Thomas Kirkman, F.R.S., Rector of Croft with Southworth, read them a paper, *On the perfect r-partitions of r^2-r+1*. Others may have known what to expect as Kirkman had already presented mathematical papers to the Literary and Philosophical Society of Manchester since 1848.

The paper on perfect partitions was published in the *Transactions of the Historical Society*, sandwiched between a paper on Zionism and one on the history of the English language – these probably being more familiar topics of interest to the Society. The paper presented a new type of combinatorial structure, now known as a planar difference set. This is related to the systems he had introduced in 1850 which are now known as finite projective planes. The new results were even less likely to be taken up by others than the earlier one, since they were both published in local journals unlikely to be read by professional mathematicians, other than those to whom Kirkman may have sent offprints.

In 1850, Cayley had written a note on the schoolgirl problem in the *Cambridge and Dublin Mathematics Journal*, in which he had raised the question of whether there was a cyclic method of generating the required triples, but doubting that this was possible in the case of fifteen elements. In devising such a method for other numbers of elements, Kirkman was led to the notion of a perfect partition of an integer.

A partition of an integer is an expression as a sum of parts – for example $1+2+4$ is a partition of 7. The classical work on the topic by Euler in the previous century had been considerably extended by Sylvester and others at about this time. In his customary eloquent style, Sylvester suggested that

> Partition constitutes the sphere in which analysis lives, moves and has its being, no power of language can exaggerate or point too forcibly the importance of this till recently almost neglected, but vast, subtle, and universally permeating, element of algebraic thought and experience.

John Herschel, the mathematician who later became a famous astronomer, had obtained in 1850 explicit formulas for $p_r(n)$, the number of partitions of n into r parts, for $r=1,2,3,4,5$. Kirkman had then derived the same formulas using more elementary methods, and had extended them to $r=6$. These were presented to the Manchester Literary and Philosophical Society on 1854 and published the following year. Sylvester had been

I tried in London to read your paper on partitions; but it beat me. As I do not subscribe now to the Cambridge Math^l Journal, I have had no opportunity of trying again. If you will send it to me & promise to answer any question that my ignorance may put to you, I will gladly give you my poor opinion. Perhaps I could make out your meaning with a little study, & I should much wish to try again.

Cayley does not write to me a single line. Did I offend him by any mode of returning to your discoveries this in my paper of the 7-partitions of x^2. I am afraid he has cut me. Accept my warm thanks for your invitation. I will certainly call on you when I come to London, & you will at any time be a most welcome guest here. Believe me, my dear Sylvester ever yours sincerely

Thos. P. Kirkman

working on the same problem and his comprehensive general account was also published in 1855. Two years later, Kirkman extended his own results to r=7 and commented that application of Sylvester's formula in particular cases was too complicated.

With this work on partitions fresh in his mind, it was natural that he should translate the patterns encountered in his cyclic generation of multiple systems into the notion of a *perfect r-partition of n* which he defined as one in which the *r* parts are ordered in such a way that every number from 1 to *n* can be uniquely expressed by one of the parts, or by sums of adjacent pairs, triples and so on, these being taken in a cyclic sense. For example, 1+2+6+4 is a 4-partition of 13 that is perfect, since every number from 1 to 13 can be expressed in the required way – thus, 5=4+1, 8=2+6, 9=1+2+6, and so on – whereas 1+1+4+4+3 is not a perfect 5-partition of 13 because the expression for, say, 4 is not unique.

Clearly the notion is a minimising one and it relates to a classical problem in recreational mathematics, dating from the seventeenth-century collection of puzzles by Bachet de Meziriac, namely that of choosing the least number of weights which can weigh any integral number of units from 1 up to say 40. Some contemporary writers define a perfect partition in this sense, but the essence of Kirkman's use of the expression is that the weights (or parts) are fixed in a cyclic order. As an illustration, a typical problem in the tradition of recreational mathematics might be to determine the number and thicknesses of a set of thickness gauges slotted on a key-ring that may be used to measure any integral thickness up to a given number. For example, four gauges of thickness 1,2,6 and 4 units, slotted in that cyclic order, would be able to measure all thicknesses from 1 to 13.

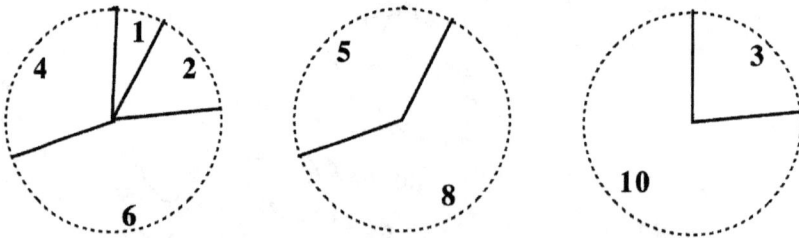

It is perhaps worth extending this digression to note that the problem becomes much more difficult when the parts remain ordered but the cyclic possibility is removed. In the linear case, an appropriate problem might be to make the least number of marks on a ruler so as to permit the measurement of all integral units of length up to a given number. When this number is 13, the measurements cannot all be done with less than 5 marks. One possibility – there are three – is that the marks are at successive steps of

1,1,4,4,3 units from one end of the ruler. In this case the partition that solves the problem is $1+1+4+4+3$ and this is not a perfect partition.

Kirkman himself related his work to the problem of numbering tickets for a foreign lottery that gave a first prize for the ticket containing all the winning numbers of a draw, but also further prizes for tickets that had all but one, or two, or three and so on, of the winning numbers.

> It was evidently of great importance to the promoters of the lottery, that they should not sell more of one combination than another, in order to equalise their own chances. They were consequently put to great expense in packing their tickets, and a great number of hands were employed by them for a length of time to make such combinations by trial as would suit the end proposed.

Kirkman showed how a perfect r-partition could be used to automatically generate multiple systems, that is systems of r-ples covering all pairs of a number of elements. Indeed this is how he had come across the notion of perfect partitions in the first place. Thus, a perfect 3-partition of 7 is $1+2+4$ and this defines what Kirkman called a *difference circle* of triples generated by adding (1,1,1) mod 7 successively to the triple (0,1,3) whose cyclic differences are the parts of the partition, namely 1,2,4 (the last being the cyclic difference 0-3 mod 7). This produces the cycle:

$(0,1,3) - (1,2,4) - (2,3,5) - (3,4,6) - (4,5,0) - (5,6,1) - (6,0,2) \ldots$

where the next step returns to (0,1,3). The cycle produces 7 triples covering all pairs of the 7 elements which have been labelled 0 to 6.

Now systems of multiples can only cover all pairs of a number of elements in certain cases. In the same way, perfect partitions are only possible for certain numbers of elements which Kirkman showed had to be of the form r^2-r+1, in which case the partitions are into r parts. He had, in 1850, already announced in a brief note extending the schoolgirl problem that there were r^2-r+1 elements with r-ples covering all pairs whenever $r-1$ is prime. His new method now yielded similar systems whenever there was a perfect r-partition of r^2-r+1. This was certainly so for $r=5$, but here $r-1$ is not prime, as his earlier result might have led him to suppose. "It was a surprise to me," he wrote, "to discover by accident a perfect partition of 21, thus proving that $r-1$ need not be a prime."

The perfect 5-partition of 21 he had found was $1+3+10+2+5$, and he showed that this was unique. It gave a difference circle of 21 elements generated from (0,1,4,14,16). Read in the reverse direction, the partition gives another circle of 21 elements generated from (0,1,6,8,18). Thus the partition yields two arrangements of 21 elements into quintuples covering all pairs, though these are structurally the same, as Kirkman showed.

For $r=6$, he found five different perfect partitions of 31:
$$1+5+12+4+7+2 = 1+7+3+2+4+14 = 1+3+6+2+5+14 =$$
$$1+2+5+4+6+13 = 1+3+2+7+8+10$$
and so derived ten arrangements of sextuples, though these also turned out to be the same.

He was unable to find a perfect partition when $r=7$, so was not able to construct a system of 7-ples covering all pairs. "I cannot demonstrate that none exists, but I think it improbable." His intuition was here very sure, since a system of r-ples derived from a perfect r-partition turned out to be a geometrical structure, now called a *finite projective plane*. This has m points lying in r-ples on m lines and it is known that there is no such plane when $r=7$.

Not all finite projective planes have the characteristic cyclic nature of the systems Kirkman derived by means of his perfect partitions. The ones that do have been called *cyclic projective planes* and it has been shown that such planes with r points on each line exist when $r-1$ is a power of a prime. It has been conjectured that this is also a necessary condition, that is that there are no others.

The first few finite projective planes up to $r=9$ are all unique and are therefore cyclic. It is interesting to recall Kirkman's surprise at his discovery of a perfect partition when $r=5$, his doubt that one could be found for $r=6$ and his further intuition that one would be found for $r=9$ (where $r-1$ is a power of a prime).

No-one ever developed Kirkman's perfect partitions any further. These involved an awkward mixture of ideas and there was no obvious general method of handling them. But as Kirkman knew, they were the key to the cyclic generation of multiple systems.

It is easy to see in hindsight that it was his difference circle that was the more fruitful and interesting notion. The parts of a perfect r-partition are cyclic first differences of every r-set in the difference circle. Each r-set provides a set of numbers whose differences yield every residue mod (r^2-r+1). For example, the quadruple $(0,1,3,9)$, derived from one of the 4-partitions of 13, provides a set of four numbers whose differences, in any order, yield all the residues mod 13. Such sets are called *planar difference sets*. They were widely studied and generalised in the nineteen-fifties, a hundred years after Kirkman's work.

Kirkman's use of multipliers to relate the r-set in the difference circles has also been found to be a key to the further analysis of difference sets. When the sets in a difference circle are multiplied by the residues mod (r^2-r+1), the result is also a difference cycle. Kirkman found that for some residues the cycle remained the same, thus creating an equivalence class of

difference sets. A natural representation for each class is the difference set
including 0 and 1. For example, Kirkman found two perfect partitions
when r=4 and so four different circles. The three multipliers 1,3 and 9,
leave each of them unchanged. There are four corresponding classes of
difference sets whose natural representations are (0,1,3,9), (0,1,4,6),
(0,1,5,11) and (0,1,8,10). All the other eighty difference sets may be considered to be equivalent to one of these.

The multipliers that define equivalence of difference sets are as complicated to deal with as perfect r-partitions, but they have attracted a lot of
study since they were re-invented by Marshall Hall in 1947.

In the introduction to his paper to the Historical Society, Kirkman
commented that combinatorial problems like that of the schoolgirls were of
high mathematical interest and – referring to his lottery example – were
"not without their practical, and even their commercial value". He also
raised an interesting more general issue about such problems:

> Their analytical interest arises hence; that it is exceedingly difficult
> to state in algebraic language the conditions of a purely tactical
> problem. The elements to be combined in these questions have no
> property except that of diversity. They have no arithmetical value or
> capacity, except that they can be counted. No operation of addition,
> subtraction, multiplication, or division can be performed upon
> them. They can merely be combined. A *tactic calculus* may one day
> be discovered, by which these problems shall be controlled; but all
> this is part of the sublime unknown. We can do nothing effectual till
> then with the analysis thus far in our hands, unless we can contrive
> to transform the *tactic* into *arithmetic*.

3

Some geometrical configurations

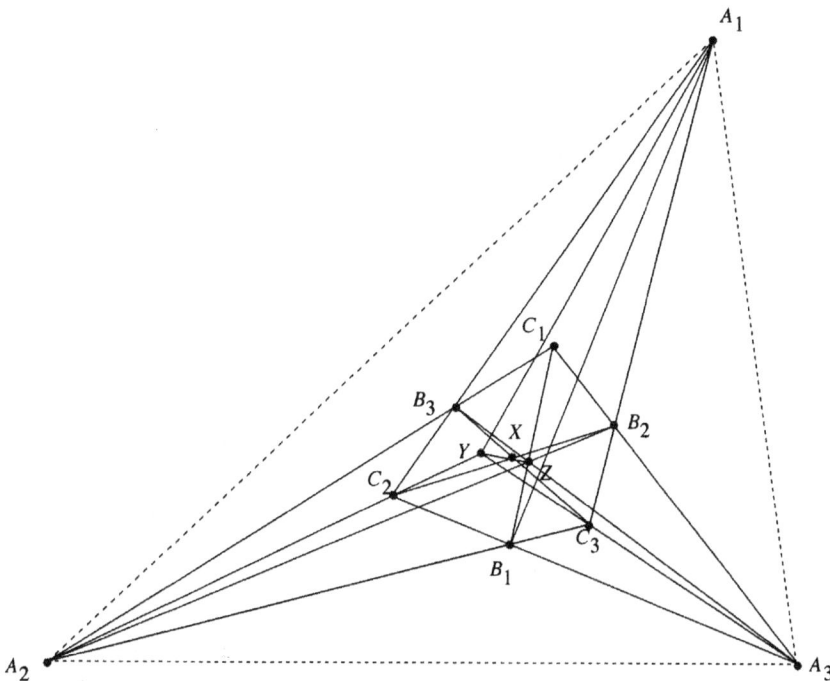

12 points lying in threes on 16 lines lying in fours on each point: a configuration formed by the trisectors of triangle $A_1A_2A_3$, forming two equilateral triangles $B_1B_2B_3$ and $C_1C_2C_3$ which are each in perspective with each other and with the original triangle on the collinear points X,Y,Z.

3.1 Porisms and Pappus

Kirkman soon turned his attention to some geometrical theorems that had a combinatorial flavour. The sort of geometry that he got involved with goes back to Euclid in the 3rd century BC – not to that famous author's *Elements*, but to another treatise, now lost, called the *Porisms*. The three books of this missing work are only known to us at second-hand, about seven centuries later, through a commentary by the Alexandrian mathematician, Pappus, in the 4th century AD.

> These porisms are in principle subtle and natural, indispensable and quite general, and afford much pleasure to those who are able to understand and investigate them.

But it is not very clear what porisms are. Pappus says that they are neither *theorems*, which required a proof, nor *problems*, which required a construction, but rather something that lay in between. It seems that a porism involved a rather general context which required further exploration, perhaps in order to then define a more precise challenge like a proof or a construction. Pappus only gave one actual example followed by some further results of his own. Many later mathematicians have attempted to reconstruct the contents of the three missing books. One of them, Robert Simson, gave an intriguing account of how he came back to the problem of a reconstruction after apparently giving it up.

> I often tried, but always in vain, to understand and restore the only porism which survives out of all that were in the three books, and as my meditations on it took up too much of my time I determined that I would never touch the subject again, especially as Halley had given up all hope of understanding it. Consequently, whenever it occurred to me, I always refused to dwell upon it. However, sometimes afterwards it presented itself to my mind when I was off my guard, and had in fact forgotten all about it, and it held possession of my thoughts until at length a glimmer of light was thrown upon it.

The surviving example of a porism described by Pappus is certainly quite a different statement than anything found in Euclid's other geometrical works. There is no way of knowing the sort of context that would have led to it.

> If, in a system of four straight lines which cut one another two and two, the three points [of intersection] of one straight line be given, while the rest except one lie on different straight lines given in

position, the remaining point also will be on a straight line given in position.

It is worth trying to unravel the meaning of this porism by drawing one's own diagrams. Perhaps the most fruitful interpretation would be a dynamic one. Thus, the situation Pappus describes could be expressed in terms of a variable triangle whose sides pass through three fixed collinear points. If two of the vertices lie on two fixed lines then the third will also lie on a line.

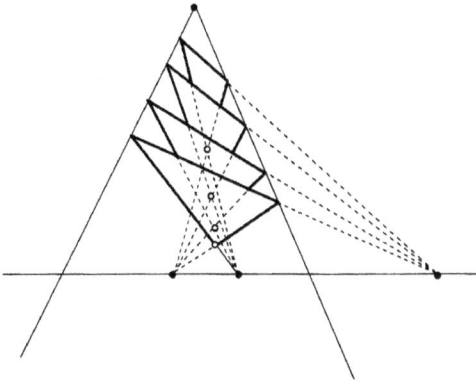

Pappus offered a rather obscure generalisation of this porism; this was one of the things Simson was able to clarify when he came back to the problem of reconstructing the porisms. The generalisation may be illustrated by taking the case of a variable quadrilateral whose sides pass through *four* collinear points. Then, if *four* of the six vertices lie on given straight lines, so in general will the remaining two.

Though the various reconstructions necessarily remain speculative, there is a general consensus that Euclid had begun to investigate notions that would now be called projective. Certainly the surviving example is not easily proved using the elementary metrical notions of the *Elements*. Moreover, the further results that Pappus provided are much more easily established in a projective context. The most important of these results is the one called Pappus' theorem and which is in fact often now taken as an axiom of projective geometry. The result may be described in various ways, but in view of later developments it is perhaps best done so in terms of a hexagon ABCDEF whose vertices lie alternately on two given lines. This means that the hexagon will necessarily cross itself. It turns out that pairs of opposite sides of the hexagon meet in three collinear points P,Q,R as shown here.

The theorem captures some fundamental property of the plane. The figure illustrating it has been called the most important figure in all geometry. This is further emphasised by the fact that what was originally found as a consequence of other assumptions, is

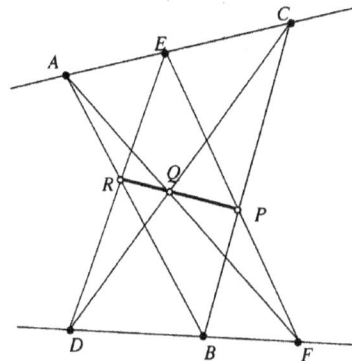

itself now taken to be a basic starting point, an axiom. Such an inversion is very characteristic of the way mathematics is continually re-organised: present-day definitions and axioms may be the hard-won and obscure theorems of previous times.

There are various different ways of looking at the Pappus property, each of which is useful when it comes to later developments. Thus, it has been introduced above as a property of a hexagon which could be said to be inscribed in a pair of lines; later it became possible to consider similar properties for hexagons inscribed in, say, a circle. Meanwhile, we could also consider a hexagon whose *sides* pass alternately through two fixed *points*, in which case the diagonals of the hexagon are concurrent. This systematically interchanges points and lines in the statement of Pappus' theorem, but as can be seen here this merely yields another way of seeing the same figure.

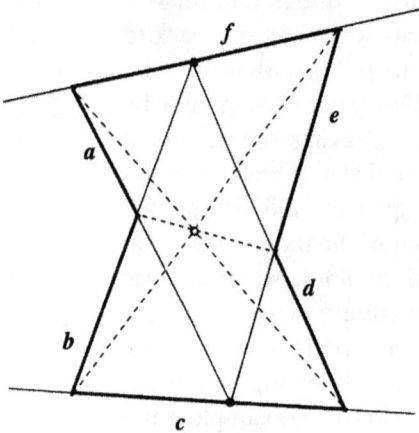

Another interpretation takes the figure as being about triangles inscribed in triangles. The triangle ABC can be seen to be inscribed in DQF, which is itself inscribed in PER, which is in turn inscribed in ABC. This yields a cycle of three triangles each inscribed in the next, and there will be various similar cycles.

Yet another way of interpreting the Pappus figure invokes the notion of triangles being in perspective. Thus, triangles APD and FRC are in perspective from Q since corresponding vertices lie on lines through Q. But these two triangles are also in perspective from B and from E. So another way of interpreting the theorem is to assert that when two triangles are in perspective from two points, they will also be in perspective from a third.

Finally, a quite different point of view indicates a connection with Kirkman's triple systems. The nine points of the Pappus figure lie in threes on nine lines which lie on threes on the points. The circular pattern of this description captures some of the combinatorial 'balance' found in a triple system formed out of nine elements. When the elements are taken to be points, the triples may be thought of as lines. The property that no pair of elements is repeated in such triples reflects the geometrical property that there can only be one line through two given points. The combinatorial structure may be illustrated by labelling the points by letters, in which case the lines of the Pappus figure may be labelled by three-letter words, as in the strange sentence: LET WAG SEW, SOP SAT LOW, LAP GOT PEG.

It is, of course, not known what sort of context Pappus had in mind for his theorem. It seems likely that Euclid must have considered something similar. But it is not clear what connection Pappus might have made between his theorem and the one original porism which he quoted. With hindsight, it is possible to look for some pattern in Euclid's porism that might correspond to the combinatorial structure of Pappus' theorem. This might be done by taking the dynamic interpretation of the porism and arresting two 'frames'. Thus, in the figure opposite the sides of triangles XYZ and X'Y'Z'

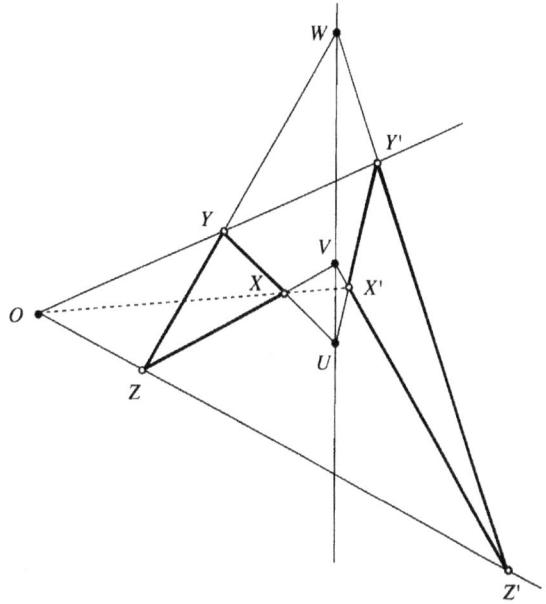

pass through the three collinear points U,V,W, and the porism asserts that if two corresponding pairs of vertices lie on a line so will the other pair.

There is already here an interesting combinatorial structure. There are nine points, each with three lines passing through it, and there are ten lines, seven of which contain three points. The exceptional lines are XX', YY', and ZZ'. If these all passed through a tenth point, the combinatorial balance would be satisfyingly complete, namely ten points lying in threes on ten lines which pass in threes through each of the points. In fact, any attempt to draw a few different figures for the porism (or to animate it on a computer screen) immediately suggests this further property, namely that the apparently exceptional lines are in fact always concurrent. Moreover, it seems a natural step to consider the position of the point X when the ray from U is drawn through the point of intersection of the fixed lines on which Y and Z lie. In this case, Y and Z will coincide at this intersection, as does X. Hence the line on which – according to the porism – X will lie, passes through this intersection.

It does seem surprising that Euclid, or Pappus, did not, as far as we know, take this step. But then nor did many later mathematicians who studied the porism. The completed figure was discovered independently some nineteen centuries after Euclid. It turned out to be another fundamental geometric property.

3.2 Desargues

The re-invention of Euclid's porism was due to Girard Desargues, an independent and original thinker who created a vast new area of mathematics that was totally ignored by most of his contemporaries.

Desargues was born at Lyons in 1591; little is known of his early life. He seems to have been at one time a military engineer; in his forties he was known in mathematical circles in Paris as a practising architect and an author of a pamphlet on perspective. In 1639, he published an extraordinary treatise on geometry in which he single-handedly created a new sort of geometry, which was not re-discovered until almost two hundred years later. A few copies of his work circulated among his friends, but these were all lost. A manuscript copy made at the end of the century was eventually discovered in a library in 1845 and an edition was printed some years later. By that time others had independently created and developed his work.

There were various reasons why Desargues' treatise did not get taken up by many mathematicians at the time. In the first place, he was an argumentative man, who followed his own bent without much reference to others. This meant, for example, that much of his work was expressed in a strange jargon, partly based on some fanciful terms used by the 15th century architect Alberti, and partly his own invention. Thus, points on a line were called *knots* of a *trunk*; segments on a line were *branches* and a point of reference on the line was called a *stump*. In effect, he created from a scratch a theory of what he called *involution* – this being the only one of his terms that has been retained.

His contemporaries were more interested in the newly developed algebraic techniques of his contemporary, René Descartes. The latter had great respect for Desargues and was a close friend, but he usually scorned any geometrical work of the traditional kind. Nevertheless the instincts of the practising architect were sound. The theory of perspective, mainly developed at first by architects and by painters, provided the appropriate continuation of the geometry that the ancient Greeks had just begun to explore.

The version of Euclid's porism now known as Desargue's theorem was a typical example of the new point of view. The theorem first appeared in an added appendix to his pamphlet which was reproduced in a book by a friar, Abraham Bosse, published in 1648. Desargues introduces a three-dimensional point of view that was quite novel for mathematicians at the time. This may be illustrated here by using – with hindsight – the same figure given with the discussion of the complete figure for Euclid's porism. The brilliant and totally original innovation was to re-interpret this diagram as being a representation in three dimensions.

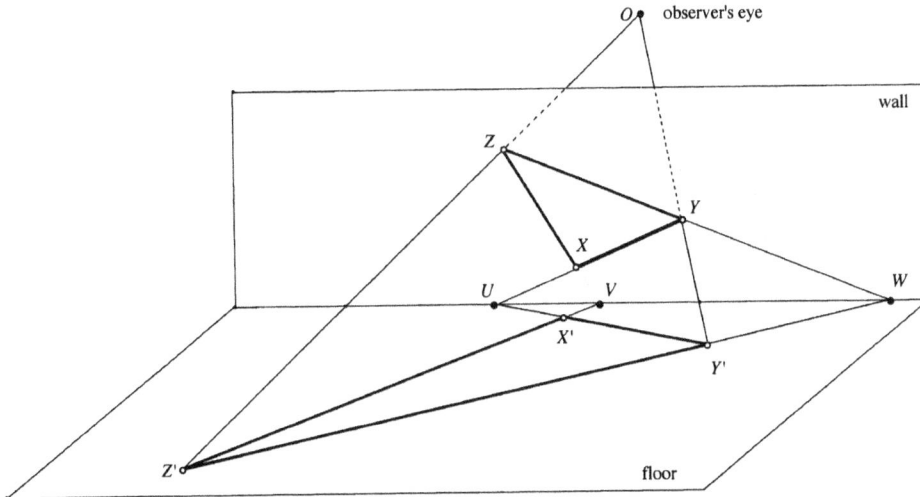

The figure is now re-interpreted as starting from the notion of two triangles being in perspective on a point. The lines through this point are interpreted as being in three dimensions – perhaps as rays from the sun or lines of sight from an eye. A triangle XYZ in one plane projects (casts a shadow) onto a triangle X'Y'Z' in another plane. Since YZ and Y'Z' lie in the plane determined by the two lines OYY' and OZZ', they will intersect in the point U, which will necessarily be on the line of intersection of the planes containing the triangles. As this is also true for V and W the required collinearity follows. The triangles may then also be described as being in perspective on a line.

This is an amazing and very satisfying proof that invokes elementary notions to prove a subtle and important result. Moreover the argument is completely reversable so that if the triangles are in perspective on a line they will be in perspective on a point. This is, of course, the dual of the original statement with points and lines interchanged.

To prove the theorem in two dimensions is much more complicated, though it is plausible to think in terms of a gradual change in the position of the two planes until they flatten into one and to argue from continuity that the collinear property remains unchanged. Alternatively, one may consider that the same diagram may represent the situation in two or three dimensions depending on a shift of attention. In this case, the representation of the collinearity in three dimensions seems to remain valid when attention shifts back to the picture plane.

Descartes did in fact himself offer a separate proof of the two dimensional case, using a classical metrical approach. But the property can be shown to be a direct consequence of Pappus' theorem and this may well be what Pappus originally intended. His theorem may have been developed as a tool with which to prove Euclid's porism. In the figure below the the theorem is applied three times to establish the Desargues property. In the first place, U,M,N are collinear since they form the Pappus line of hexagon OKY'XYZ; then VNL is the Pappus line of OZ'X'Y'XK; and finally UVW is the Pappus line of LY'KXMN, and the triangles are in perspective on this line as required.

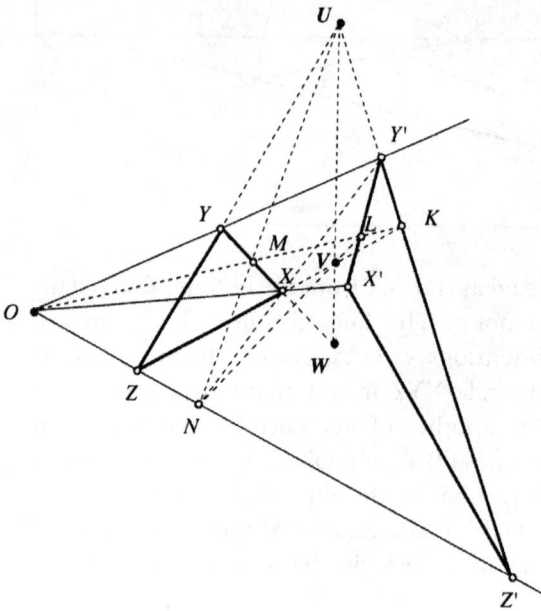

As with the Pappus figure, it is interesting and fruitful to interpret the Desargues figure in different ways. Thus, any one of the ten points can be considered to be a centre of perspective so that there are ten ways of interpreting the figure as a pair of triangles in perspective. In the figure on the right two triangles are shown in perspective on the point X' and on the line UYZ.

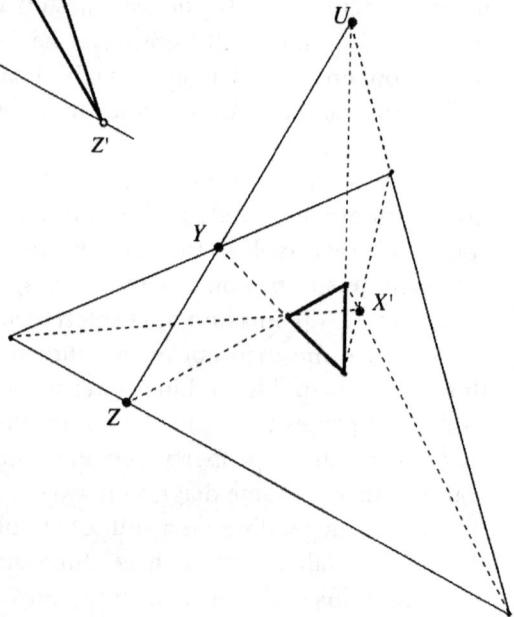

Another way of looking at the figure is to see the ten lines as two mutually inscribed pentagons; this can be done in various different ways.

Finally, the figure can be interpreted as a combinatorial structure, ie a triple system formed from ten elements – for example, by the letters and words of the sentence: NOW SHE WET HUN PEN WHY YOU SUP POT STY.

The Desargues figure is so fundamental that it is often nowadays taken to be an axiom.

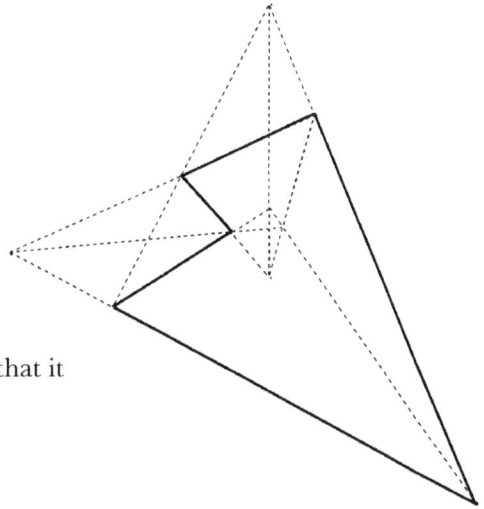

3.3 The Pascal configuration

The philosopher and mathematician Blaise Pascal had an unusual education. His father had not wanted his son to study mathematics before the age of fifteen and would not let any mathematics texts in the house. This made the young boy curious; he explored geometry for himself. The father relented, let him read Euclid and then took the fourteen-year-old with him to the meetings that used to be convened by the friar Marin Mersenne for a small group of mathematicians. Pascal met Desargues at these meetings and came to appreciate his work. Later he was to write that Desargues was of one of the great geniuses of his time: "I should like to say I owe the little I have found on this subject to his writings and that I have tried to imitate his method as far as possible".

In 1639, when he was sixteen, Pascal presented Mersenne's seminar with a list of theorems which included the statement that the meets of opposite sides of a hexagon inscribed in a circle are collinear. The projective link between a circle and conics in general was well understood and the property would have been understood to apply to hexagons inscribed in a conic. With hindsight, this may be seen to include Pappus' theorem as the special case where the conic is a pair of lines, though there is no evidence that this link was made at the time.

It is not clear how Pascal might have proved his theorem. There have of course been many subsequent proofs of what turned out to be an important theorem of projective geometry. But what is particularly interesting in the present context is the neat algebraic proof offered by Kirkman in a 1849 paper on hexagons inscribed in a conic. This uses an abridged notation for lines that had some years earlier been succesfully introduced by Gabriel Lamé and later developed by Etienne Bobillier. The linear expressions for the equations for the sides of the hexagon are represented by single letters a,b,c,d,e,f. Kirkman used another notation, but in effect formed an equation $a.c.e - K.d.f.b = 0$, where K is some constant. This is of the third degree and so represents a cubic curve passing through the intersections of opposite sides a,d, and so on. It also passes though the vertices which are intersections of consecutive sides a,b, and so on. K is then chosen so that the curve goes through some seventh point of the conic. But since a general cubic can only meet a conic in six points, this one must break up into the conic itself and a straight line, which will be the required Pascal line.

Kirkman's paper was first published in 1849 in two issues of the *Manchester Courier*, a Tory local newspaper founded in 1825, a few years after the liberal *Manchester Guardian*. It is not clear how many of its readers would have understood his highly technical article, or indeed whether it was read by the mathematicians who could, but it was reprinted the following year in the *Cambridge and Dublin Mathematical Journal*.

This was a busy and rewarding time for Kirkman. After he moved to Croft in 1845, he had begun to work at different areas of mathematics that caught his interest. His first paper on combinatorics in 1847 was followed in the following year by the paper on what he called pluquaternions. It is perhaps not surprising that he then turned his attention to the combinatorial structures involving Pascal lines which had been assiduously explored, particularly by German geometers, in the preceeding decades. In his paper on what he called the complete hexagon, Kirkman reviewed some of results and established some of his own.

Six points on a conic can be joined up to form a hexagon from a chosen point in $5.4.3.2.1 = 120$ different ways, but this counts hexagons in reverse order, so that there are but 60 distinct Pascal lines. There are 15 chords which intersect in 45 points; and each of these chord-intersections lie in 3s on the Pascal lines (for example in Kirkman's algebraic treatment mentioned above, the Pascal line contains the three intersections of a with d, b, and f). In turn, the Pascal lines lie in 4s on the chord-intersections. This establishes a particular instance, denoted by (60,3; 45,4), of a geometrical figure $(x,y; X,Y)$ in which x points lie in ys on X lines which lie in Ys on the points. The total number of point-line incidences is $x.Y$ or equally $X.y$, so that the figure has a sort of combinatorial balance expressed by the equation $x/y = X/Y$.

It had already been discovered by Jakob Steiner that the Pascal lines also lay in 3s on 20 points, which were subsequently named after him. In the figure below, the Pascal lines of hexagons ABCDEF, ADCFEB, AFCBED inscribed in a conic intersect in a Steiner point S.

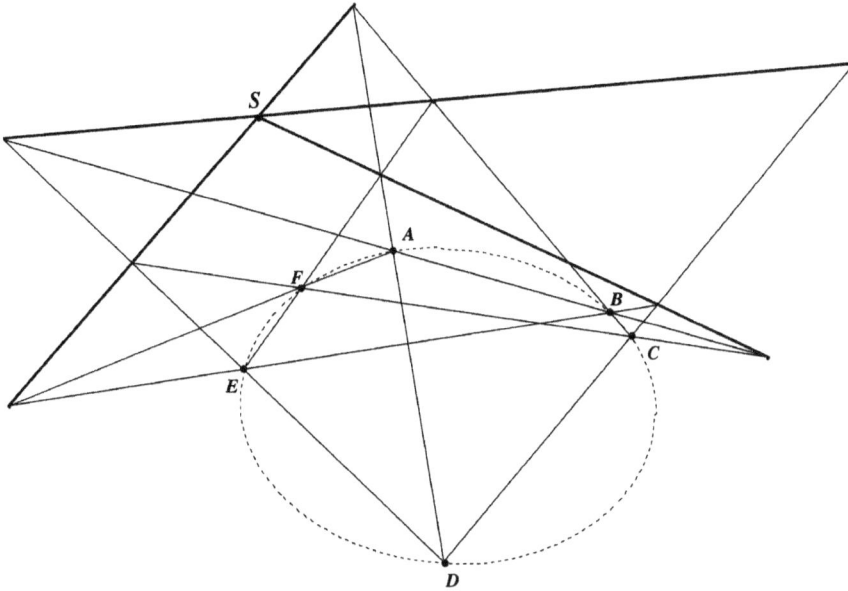

Steiner had also mistakenly asserted that his 20 points lay in 4s on 5 concurrent lines, but this had been corrected by his great rival Julius Plücker who showed that the 20 Steiner points lay in 4s in fact on 15 lines (now named after him) which lay in 3s on the points. This may be drawn as a geometric figure, but as with all such configurations it could be represented by 20 letters forming 15 four-letter words:

MANY SKIP FEST FIND GHAT HUMP FLOG BRIG
BUNK COSH CURT PREY LACK BODY MELD.

Kirkman used his condensed algebraic notation to prove these results, which he carefully ascribed to the original authors. But he also added a new property of his own. He showed that there were 60 points which lay in 3s on the 60 Pascal lines which themselves lay in 3s on the points. In the figure overleaf, the Pascal lines of hexagons ADFBEC, AEBDFC, and AECFBD inscribed in a conic intersect in a Kirkman point.

This meant that each Pascal line had not only a Steiner point but also three of what were then called Kirkman points. The resulting figure (60,3; 60;3) establishes a sort of dual correspondence between Pascal lines and Kirkman points which extends to the configurations containing them. Thus, corresponding to the Steiner intersections of Pascal lines, namely (20,1; 60,3) Arthur Cayley proved the corresponding (60,3; 20,1) in which the 60 Kirkman points lie in 3s on 20 lines, which then became known as Cayley lines. This result is ascribed to Cayley in Kirkman's paper, at any rate

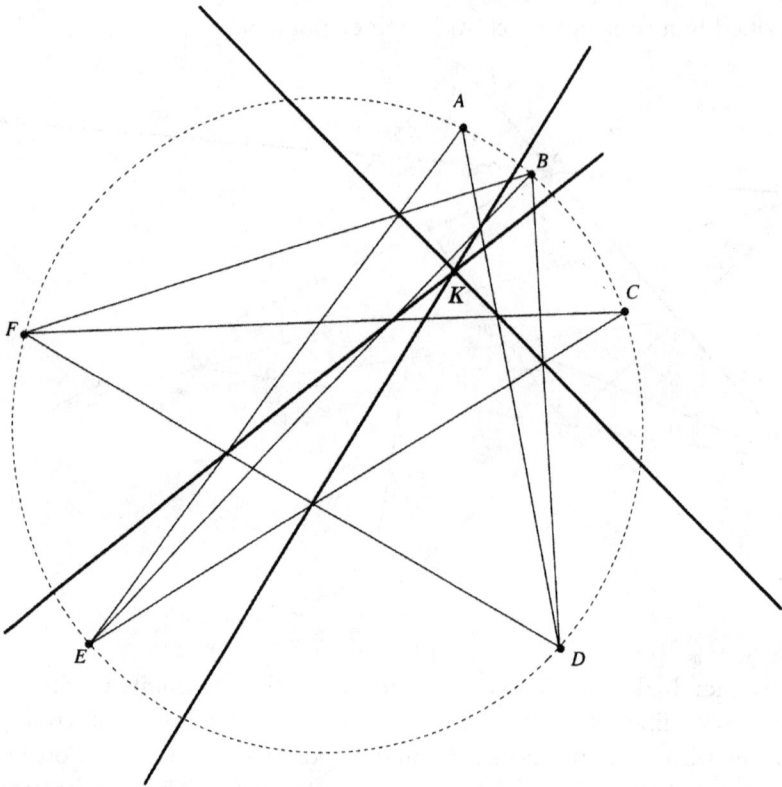

in the version of it that was re-printed in 1850, where he writes that he was offering a different proof than the one Cayley had sent him. It is known that he had been in correspondence with Cayley and it is likely that he had communicated his work on Pascal lines when he had been working on it earlier in 1849.

It is also known that he exchanged letters at this time with George Salmon, whose later text-books on conic sections were to make the Pascal line configurations more widely known. It was Salmon who found the figure corresponding to the Plücker collinearities of Steiner points, namely (20,4; 15,3), This was (15,3; 20,4) in which 15 points, subsequently known as Salmon points, lie in 3s on 20 Cayley lines. One representation could be:

MOB HOG TOP SOL LAB HER HUM HAS RUN UPS
ANT BEG ELM RIM LIP PEN NIB GUT ITS RAG

This result was also given in Kirkman's article where he acknowledges that the proof he provides was due to Salmon. He also goes on to develop a number of complicated further results; for example, that the 45 chord-intersections lie in 6s on 60 conics. Kirkman comments:

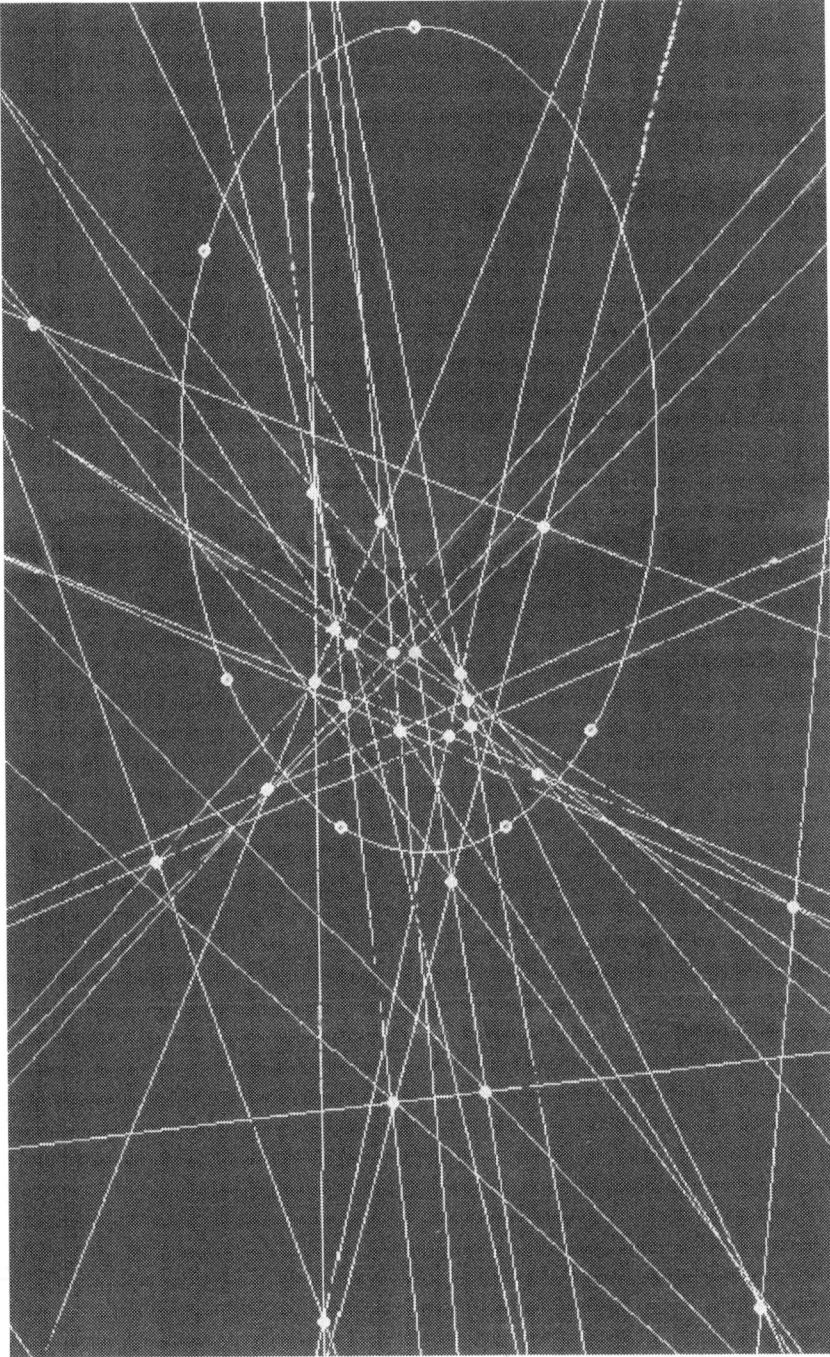

Figure by Steve Sigur of the 60 Pascal lines of a hexagon inscribed in a conic with some of their intersections at Kirkman points

> It is easy to find thousands of conics, each fulfilling six or more
> conditions at the intersections of the leading lines of the figure; and
> curves of higher degree, which fulfil remarkable conditions in the
> figure, amount to an enormous number.

He ends his paper with the remark that his results also have corresponding
duals. Curiously enough the actual dual of Pascal's theorem of 1639 – that
the joins of opposite vertices of a hexagon whose sides touch a conic – had
apparently not been considered until 1808 when it was first formulated and
proved by Charles-Julien Brianchon.

In a final looking back to Desargues and his development of the work of
Euclid and Pappus, it may be noted that the 60 Kirkman points break up
into six Desargues figures (10,3; 10,3). One of these is shown below.

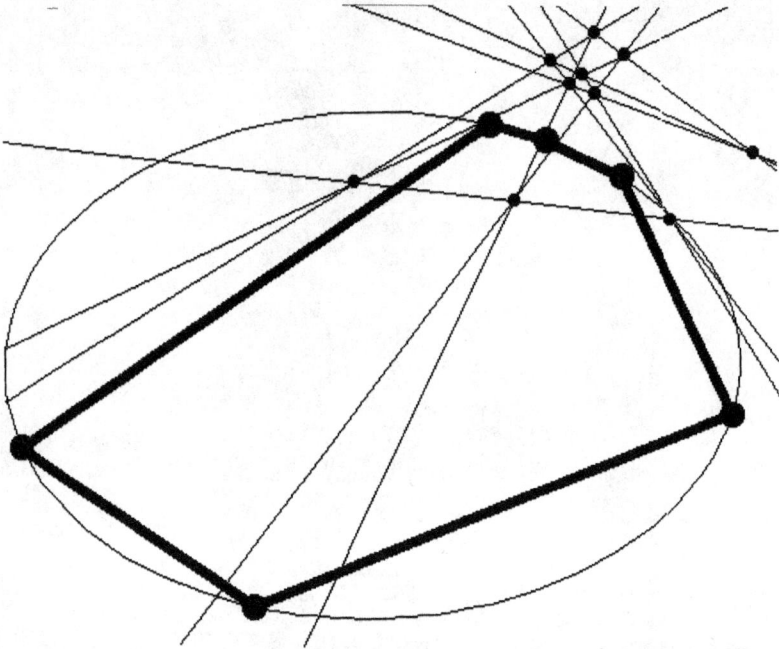

Kirkman followed his paper on the Pascal configuration a few months
later with another paper on algebraic geometry, *On the meaning of the
equation $U^2 = V^2$ where U and V are products of n linear forms of two variables*,
which included some theorems on curves of general degree from which he
derived new proofs of Pascal's theorem for conics. He mentioned that after
discovering these he found that they were implied by a theorem of Plucker's
and he gave a specific reference.

1850 was a prolific year for Kirkman. Apart from these papers on
geometry, he also prepared two further papers on combinatorial systems in

August and one on a development of quaternions in September. He never specifically related his work on the Pascal configuration or on quaternions to his studies of combinatorial systems and it is very likely that these various strands of interest were independently motivated. But a common thread is easily now detected.

All the above papers were in print before the end of the year. The appearance of many new specialised scientific journals at the beginning of the nineteenth century – in England, for example, the *London, Edinburgh and Dublin Philosophical Magazine* from 1798, the *Proceedings of the Royal Society of Edinburgh* from 1832, the *Cambridge and Dublin Mathematical Journal* from 1837, the *Proceedings of the Cambridge Philosophical Society* from 1843 – was an indication of the increasing number of professional mathematicians. Nevertheless there were not so many of these that an amateur like Kirkman could not get his papers published quickly.

Another paper on algebraic geometry appeared the following year. Pascal's theorem establishes a general relation among six points on a conic. In general, five points determine a conic and the theorem provides a construction for a sixth point. Kirkman, like many other mathematicians, was also interested at about this time in the extended problem of finding a geometrical relation among ten general points on a second degree surface. In this case, nine points in general determine such a surface and the required relation would be equivalent to the construction of a tenth point. Kirkman generalised the problem to higher dimensions in a paper, *On linear constructions*, published in 1851. This was one of the earliest investigations in 'space' of more than three dimensions and Kirkman threw in a typical aside for any reader who may have felt startled by this development.

> If the reader feel distressed with the effort to imagine such transcendental volumes in space of more than three dimensions, that is no affair of mine; my duty being, not to supply him with additional senses, but with sound arguments, of which he is competent to judge with even fewer than five. Had the reader been so unhappy as to enter this world deprived of touch, he would probably have been as much in the dark about the geometrical import of x,y,z as he now is, being endowed with only five senses, concerning the real existence of these solids of fifteen dimensions.

In this paper, Kirkman applied his general results to another particular construction problem, namely that of constructing the ninth intersection of two cubic curves through eight given points. This had been proposed as a prize question by the Brussels Academy some years before. A solution had been proposed in 1850 by Thomas Weddle, a mathematician at the Royal

Military College at Sandhurst, but this had involved a compass construction and Kirkman offered a construction by ruler alone – this could involve up to 150 applications of the ruler. A solution was also published at the same time by Andrew Hart, one which Kirkman wrote, "may be pronounced perfect, and which for elegance and simplicity leaves nothing to be desired".

3.4 Configurations, allineations and the planting of trees

Various systems of points, lines or planes were assiduously investigated in the nineteenth century. The earliest examples have already been mentioned. Thus in 1828, Steiner identified 15 points lying in 3s on 20 Pascal lines, which intersect at the points in 4s. Cayley showed in 1849 that there are 27 lines on a cubic surface lying in 3s on 45 planes, which intersected on the lines in 3s. These – like the increasingly complicated other examples that would be developed – are geometrical figures with a particular balance in that they are systems of multiples that cover each element a fixed number of time, though they do not necessarily cover all pairs of elements. The figures were called *configurations* by Karl Reye in 1876. They were considered for some time to be the most important unifying topic in geometry, and though this is no longer the case, some of the key structures in geometry were identified through the investigation of configurations.

A configuration of points and lines may be described as a system of x points lying in y-sets on X lines which cover each points Y times. Counting the repetition of points in two ways gives a necessary condition on the parameters, namely $x.Y = X.y$. Counting the number of joins of a particular point with others gives a necessary inequality $Y \leq (x-1)/(y-1)$. When the equality holds, the configuration is line-complete, in the sense that the lines now cover all pairs of points and the system is in fact a particular case of a more general combinatorial structure that later came to be called a block design (this is briefly discussed in section 5.1).

For every configuration of points and lines there will be a dual configuration with the roles of points and lines interchanged. For example, in the trivial case when $y=1$ and $X=x.Y$, all configurations are disconnected repetitions of y lines through a point; the dual configurations have $Y=1$, $x=X.y$ and are disconnected repetitions of y points on a line.

A self-dual configuration will have the same number of points as lines and so the same coverings, so that such configurations have $x=X$ and $y=Y$ and may be denoted by the symbol x_y. The necessary inequality for the parameters here means that these can only exist when $x \geq y^2 - y + 1$. For example, when $y=2$, the self-dual configurations x_2 are the x-gons, that is triangles, quadrilaterals, pentagons and so on.

All other configurations with $y=2$ may be realised as x-gons with certain diagonals drawn in, though the internal intersections of the diagonals do not count as points of the configuration. Thus a quadrilateral with its diagonals, but not their point of intersection, is a configuration with $x=4$, $y=2$, $X=6$, $Y=3$ and is represented in the plane by a (point-) complete quadrilateral, and in three-dimensional space by the faces and edges of a tetrahedron. More generally, a configuration with $y=2$ and any Y is a x-gon, $x \geq y+1$, with $y-2$ diagonals at each point drawn in. The duals of these are represented for $y \geq 4$ by the star shapes formed by extending the sides of the polygons (these are sometimes called polygrams) with some 'diagonal' intersection points on each side being counted as parts of the system. The dual of a (line-) complete polygon is a (point-) complete polygram.

The inventory of configurations becomes more interesting when $y=3$. The cases when $Y=1$ or 2 have already been mentioned as duals. When $Y=3$, the configurations are self-dual x_3 where x cannot be less than 7. Any 7_3 configuration would be complete and so a triple system in Kirkman's sense. But such a system exists and can be represented algebraically by letters; for example, by those of the words: ADO, ORE, BAR, BOY, YEA, BED, DRY. This cannot be constructed with ordinary points and lines: the seven points could be placed at the vertices, midpoints of the sides, and centroid of a triangle, and this yields six lines each containing three points; but the required balance needs an impossible seventh line through the midpoints. That a line cannot be drawn to intersect all three sides of a triangle is a seemingly obvious property of ordinary space, yet it would be difficult to prove. In a rigorous treatment of euclidean geometry this property, or something like it, would have to be assumed, and this was done in the axiomatic treatments of geometry at the end of the nineteenth century.

On the other hand, the existence of a triple system of seven elements suggests that there are other 'spaces' in which a 7_3 configuration can be constructed (the example of the Fano plane has been discussed in section 2.2). After the development of non-euclidean geometries, mathematicians found it natural to extend the use of geometrical language to other situations. Thus there is certainly a triple system of seven elements and if it cannot be constructed in ordinary geometry, then it can itself yield a new geometry. In this sense, there is a 7_3 configuration after all. Its seven points and seven lines constitute a finite geometry. So do all the other multiple systems with y^2-y+1 elements, originally studied by Kirkman and later called finite projective planes.

The next case in an inventory of configurations is 8_3. Here again there is in the abstract a set of letters with the required properties: FIE, REG, FOR, AIR, FAN, ONE, GIN, AGO. But this cannot be constructed as a configuration of ordinary points and lines. Augustus Möbius came across this system when investigating the points of inflexion of a cubic curve. He found that any eight of these nine points formed an 8_3 configuration, but since at least five of the eight points are imaginary the configuration could only be constructed in the complex plane. It is interesting to note that although Möbius was quite at ease with the notion of a geometry of complex points, he was very cautious about a space of four or more dimensions. When remarking that mirror-reflections in three dimensions might be superimposed on each other in four dimensions, he added, "since however such a space cannot be thought about, the superposition is impossible."

The 7_3 and 8_3 configurations are unique in the sense that both of them lead to new geometries. But in the case of 9_3 there are three different systems with the required properties, each of which can in fact be constructed with ordinary points and lines. The most important of these – and according to David Hilbert the most important configuration in all geometry – is the one known as the Pappus configuration, which can be represented by the sides of a hexagon whose vertices lie alternately on two lines. The intersection of 'opposite' sides of the hexagon lie on a ninth line completing the configuration. The fact that such a ninth line always exists is an important geometrical property that may be deduced from other assumption, or taken as an axiom to define a certain sort of geometry.

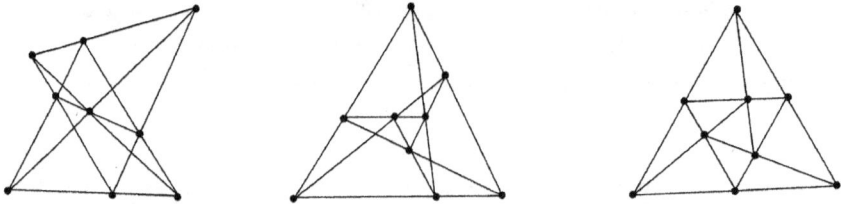

The other two 9_3 configurations do not get automatically completed when being drawn, as is the case for the Pappus configuration. In other words, they reflect the characteristics of a special arrangements of points, rather than a general property of the plane. This is also the case for eight of the ten different 10_3 configurations. Another cannot be constructed with ordinary points and lines. But there is one, known as the Desargues configuration, which like the Pappus configuration expresses an important geometrical property. (These two configurations are discussed in more detail in sections 3.1 and 3.2.)

The list of x_3 configurations continues with 31 different forms of 11_3 and 288 forms of 12_3. After these and further x_3 configurations, the inventory would continue with configurations having $y \geq 4$, but no more will be listed here.

The most important of the elementary configurations are the Fano 7_3, the Pappus 9_3 and the Desargues 10_3. These are fundamental in the sense that they may be used to characterise certain sorts of geometry. The systematic study of these and other configurations towards the end of the nineteenth century enabled mathematicians to define the combinatorial characteristics that are specifically geometrical, indeed to specify in very general terms what geometry is.

Meanwhile, whereas many configurations could not be constructed with ordinary points and lines and so led to the creation of new geometries, there remained the question of what part of them could be so realised. For example, the Fano 7_3 cannot be drawn in an ordinary plane, but the seven points and six of the lines can. Such an arrangements of seven points lying in collinear triples is the largest part of the 7_3 configuration that can be drawn with ordinary points and lines. In general, any configuration will have a number of 'lines' $X = xY/y$ which, since $Y \leq (x-1)/(y-1)$ as was shown above, will be at most $x(x-1)/y(y-1)$. This maximum value for X is only attained by the so-called complete configurations. Now the only complete configuration that can be drawn with ordinary points and lines are the complete polygons in the case $y = 2$. When $y = 3$, the complete configurations are Kirkman triple systems and these cannot be constructed with ordinary points and lines. But they set an upper bound, namely $x(x-1)/6$, for the number of lines in any configuration with $y = 3$ and so in any actual arrangements of x points lying in collinear triples.

The determination of the 'real' part of a configuration can be compared to the determination of the real part of the solution of an algebraic equation. So that it is interesting to note that Sylvester who had investigated the latter problem in 1865 turned his attention two years later to the problem of arranging a given number of points in as many collinear triples as possible. Such arrangements will not be configurations since there will not necessarily be the same number of lines through each point, but the number of lines will be bounded by the number permitted in a configuration. Thus for x points there certainly cannot be more than $x(x-1)/6$ collinear triples. This number will never actually be achieved – the determination of the actual best possibe result turns out to be a very difficult, and still unresolved, problem.

Sylvester found a general method of constructing points on a cubic curve that lay in collinear triples and set a problem on this to readers of the

Educational Times for April, 1867. He presented this more picturesquely in the September issue:

> Show ... how to plant 81 trees so as to form 800 rows of 3 trees in a row; and show also how to plant 80 trees so as to form 799 rows of 3 trees in a row.

He gave a solution in the December issue and also offered a general formula for the number – of what he called allineations – of triples of x points. His construction of triples of x points on a cubic yielded $[(x-1)^2/8]$ allineations or lines (the square brackets here denoting the largest integer less than or equal to the expression inside). This gave the 800 rows for 81 trees and the 799 rows for 80 trees.

An improved construction was suggested by Sylvester in the *Educational Times* for February, 1868, and this increased the number of allineations to $[(x-1)(x-2)/6]$, He asserted without explanation or example that these results could be "most probably" improved even further. This is certainly true since for $x=9$ his new method gave 9 allineations and this is achieved by the Pappus configuration which in a symmetrical version yields 10 allineations.

Sylvester explained his interest in allineations during an address to the British Association for the Advancement of Science on the occasion of its annual meeting in Plymouth in 1869.

> So I found as a matter of observation, that allineation in ornamental gardening – ie a method of putting trees in positions to form a very great number, or the greatest number possible, of straight rows, of which a few special cases only had been previously considered as detached porismatic problem – form part of a great connected theory of the pluperfect points on a cubic curve, those points of which the nine points of inflexion and Plucker's twenty-seven points may be seen as the lowest instances.

As Sylvester noted, occasional problems about the arrangement of trees were fairly common in the various journals, like the *Educational Times*, that set mathematical puzzles for their readers. They date back to at least 1821 when a book called *Rational amusements for winter evenings* included ten tree-planting problems – for example:

> Your aid I want, nine trees to place
> In rows just half-a-score,
> And let there be in each row three.
> Solve this: I ask no more.

The solution offered was the one provided by the symmetric version of the

Pappus configuration. Its 10 lines is still the best possible solution for 9 points. It has been claimed that this was originally found by Newton and if this is the case it is worth recalling that another seventeenth-century author, Sir Thomas Browne, also wrote about the arrangements of trees in *The Garden of Cyrus*, published in 1658. Cyrus, a famous Persian emperor, was described as being brought up in the woods and mountains:

> When time and power enabled [he] pursued the dictate of his education, and brought the treasures of the field into rule and circumspection … Not only a lord of gardens, but a manual planter thereof: disposing his trees, like his armies in regular ordination.

It was typical of Sylvester's unifying and cultured mind that he should seek to relate a general mathematical interest in the configurations of algebraic geometry to a particular problem in mathematical recreations. Like Cyrus, a mathematician always seeks to bring the treasures of the field into rule and circumspection. Sylvester obviously remained fascinated by the problem of trying to increase the number of allineations of a given number of points. It seems natural do this by joining all pairs of points – in other words to try and draw an arrangement of points and lines that form a triple system, or in yet further words to draw a complete configuration with $y=3$. Sylvester conjectured that this was not possible, even when there could be more than three points on a line, unless the points were already on a single line or were infinite in number, in which case a possible arrangement would be the vertices of an equilateral grid. He could not prove this and he presented it as a problem in the *Educational Times* in 1892:

> Prove that it is not possible to arrange any finite number of real points so that a right line through every two of them shall pass through a third, unless they all lie in the same straight line.

This does not restrict the lines to contain just three points. There may be any number greater than three of points on a line, but the conjecture asserts that there is no way in which all pairs of a finite number of points are covered unless the points are all collinear to start with. No solutions were offered by the readers of the *Educational Times* and the conjecture remained unresolved until it was proved by Nicolaas de Bruijn and Paul Erdos in 1948.

This fascinating result characterises a finite set of points. It has been interestingly re-phrased by Theodore Motzkin: for any finite number of non-collinear points there is at least one straight line containing just two points. This was improved by Leroy Kelly and William Moser in 1958 by the assertion that for n non-collinear points there are at least $3n/7$ straight lines containing just two points.

It was inevitable that the three-in-a-row tree planting problem would be extended and generalised. Four-in-a-row arrangements were posed by Ernest Dudeney (as given below) and five-in-a-row arrangements by William Rouse Ball in books on mathematical recreations. But there are few general results in these cases. As Dudeney wrote:

> Those tree-planting puzzles have always been a matter of great perplexity. They are real puzzles in the truest sense of the word, because nobody has yet succeeded in finding a direct and certain way of solving them. They demand the exercise of sagacity, ingenuity and patience, and what we call luck is also of some service. Perhaps some day a genius will discover the key to the whole mystery.

In his *Canterbury Puzzles*, Dudeney had a ploughman referring to a plantation of 16 trees set out in 12 rows of four (one of of twelve possible examples is show below). He reports that "a man of deep learning" passing through the district said that they could be planted in 15 rows of four. "Can you show me how this can be?" Dudeney claimed that his solution (shown below) was the best possible.

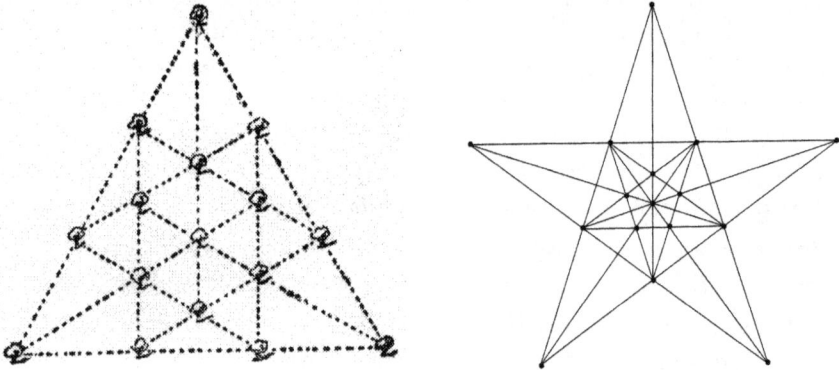

As far as the original three-in-a-row problem is concerned, the greatest number of allineations has only been established for $x \leq 12$ points. For any x the number must be at least that provided by Sylvester's construction, namely $[(x^2-3x+2)/6]$. It cannot be more than $(x^2-x)/6$ as has been shown, but this bound can be reduced in some cases by using the formula given by Kirkman in 1847 for incomplete triple systems. Thus for $x=0$ or 2, mod 6, a system of triples covering all pairs is not possible, but the largest incomplete system has $(x^2-2x)/6$ triples and this is a better upper bound in this case. It has been conjectured that except for $x=7,11,16,19$ the greatest number of allineations is $[(x^2-3x+6)/6]$. This formula fits all the known cases, but some of them have not yet been proved to be the best possible.

The solutions of Sylvester's allineation problem can now be seen as 'realisations' of Kirkman's incomplete triple systems with ordinary points and lines. There is no evidence that either of the two men observed this, but they were a little touchy about each other's work – at any rate in the area of recreational mathematics – and may have preferred to cover up any trails they had made to or from each other's work.

Inventory of configurations (m,n,h,k) with $m/n=h/k$, namely arrangements of m points on n lines with h points on each line and k lines through each point.

h=1

$(n,nk,1,k) = n$ x $(1,k,1,k) = n$ lots of k lines through 1 point

h=2

$(2n,n,2,1) = n$ x $(2,1,2,1) = n$ lots of 1 line through 2 points

$(n,n,2,2) = n$-gon

$(2n,3n,2,3)$ $n\geq2$ = $2n$-gon with diameters

$(n,2n,2,4)$ $n\geq3$ = n-gon with 2 diagonals at each vertex

....

$(n, n(n-1)/2, 2, n-1)$ = complete n-gon

h=3

$(3n,n,3,1) = n$ x $(3,1,3,1) = n$ lots of 1 line through 3 points

$(3n,2n,3,2)$ $n\geq3$ = $3n$ points of $n\times n$ grid; $n = 2$, quadrilateral

$(n,n,3,3)$ $n\geq7$ = Fano ($n=7$), Möbius (8), Pappus (9), Desargues (10)

....

$(3n,4n,3,4)$ $n\geq3$ = ... Salmon-Cayley (5) .... Pascal lines (15)

....

4

Prize questions on polyhedra and on groups

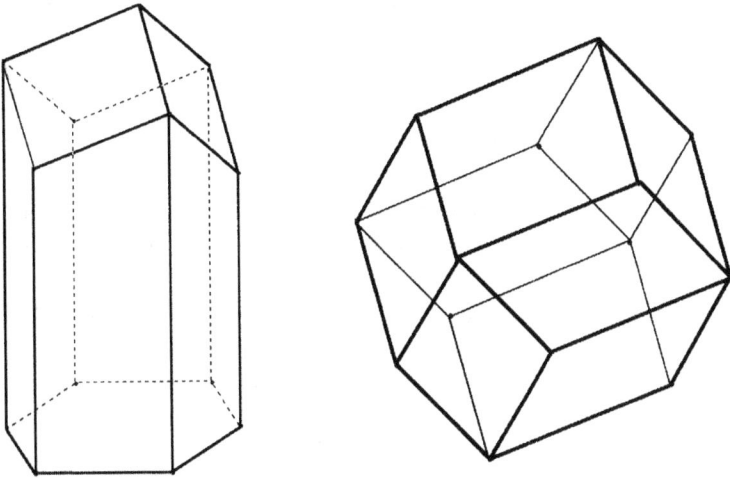

If we experiment, not with cylinders but with spheres, if for instance we pile bread-pills together and then submit the whole to a uniform pressure, as we shall presently find that Buffon did: each ball (like the seeds in a pomegranate, as Kepler said) will be in contact with twelve others – six in its own plane, three below and three above, and under compression it will develop twelve plane surfaces. It will repeat, above and below, the conditions to which the bee's cell is subject at one end only; and since the sphere is symmetrically situated towards its neighbours on all sides, it follows that the twelve plane sides to which its surface has been reduced will be all similar, equal and similarly situated. Moreover, since we have produced this result by squeezing our original spheres close together, it is evident that the bodies so formed completely fill the space. The regular solid which fulfills all these conditions is the rhombic dodecahedron. The bee's cell is this figure incompletely formed; it represents, so to speak, one half of that figure, with its apex and the six adjacent corners proper to the rhombic dodecahedron, but six sides continued, as a hexagonal prism, to an open and unfinished end.

(from D'Arcy Wentworth Thompson, *On Growth and Form*, Ch VII.)

4.1 Polyhedra

In the 1850s, Kirkman wrote a number of papers on polyhedra (which he always insisted in spelling without the 'h', claiming that people were divided on the issue and did not in any case write 'perihodic'). He was interested in classifying polyhedra with three faces at each vertex (ie trivalent polyhedra, such as the cube or tetrahedron). His first paper, read to the Manchester Literary and Philosophical Society in 1853, listed all trivalent octahedra. Further papers, submitted to the Royal Society, continued the classification to polyhedra with 10, 11 and 12 faces.

One of the latter papers also considered some possible closed polygons formed by edges of a polyhedron that included each vertex once and only once. Kirkman showed that there could be no such polygon for polyhedra with an odd number of vertices and even numbers of edges to each face. In contemporary terms, this might be expressed by noting that the graph of a polyhedron with even-edged faces is bipartite (ie its vertices fall into two disjoint sets such that no vertex in one is adjacent to one in the other), and then, that a bipartite graph with odd vertices cannot have what would now be called a circuit.

Kirkman gave the example of a bee's cell. This is in effect a distorted hollow hexagonal prism with one end sealed off by three rhombuses (as shown on the previous page). According to D'Arcy Thompson this shape has "attracted the attention and admiration of mathematicians from time immemorial". He quotes Ausonius writing of the *geometrica forma favorum* and Pliny noting that men gave a lifetime to its study. It was Kepler who first noted that the shape was that of a rhombic dodecahedron with one corner and its three rhombi sliced off. And it was René Réaumur who claimed that the shape gave the least surface area for a given volume – it held the most honey for the least wax (though this was not quite true).

For Kirkman, the bee's cell was a 13-acron (ie it had 13 vertices) with 9 quadrilateral faces and 1 hexagonal face. As such it had no closed polygon, or what would now be called a Hamiltonian circuit. Hamilton had introduced the notion in terms of a game he had devised that involved tracing a circuit on a dodecahedron; but this was in 1856, a year after Kirkman had considered the issue for general polyhedra. A version of the Icosian Game was marketed in 1859; this was named after the twenty numbered pegs that had to be appropriately inserted in holes representing the vertices of a dodecahedron. The two mathematicians corresponded from time to time and it is likely that they knew of each other's work on circuits. In one letter Kirkman expressed the wish that he had "the good fortune to be nearer to

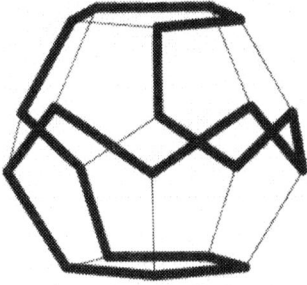

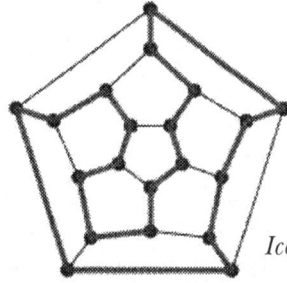

Icosian game

such a mathematician as you". And Hamilton replied that "it would be difficult for me to express, without having the air of flattery, how much I admire
your mathematical genius and discoveries". He is known to have called on
Kirkman in 1861 after the meeting of the British Asssociation in Manchester
that year. Twenty years later, Kirkman claimed in one of his contributions to
the *Educational Times* that he been the first to consider circuits in general as
well as on a dodecahedron.

Kirkman continued his work on polyhedra in the 1850s, though his
papers became more and more obscure. One in the *Proceedings of the Royal
Society* for 1857 had to have a note by Cayley introducing "this investigation
of a subject so new and intricate". He was elected a Fellow of the Royal
Society that year, this being mainly for his work on quaternions and partitions of numbers. Then, by an extraordinary coincidence, Kirkman read in
a French journal in the following year that the Prize Question being
proposed by the Académie des Sciences was to "perfectionner en quelque
point important la théorie géométrique des polyhèdres". The prize of three
hundred francs was to be awarded three years later; it was of course the
chance to establish an international reputation, rather than the money, that
spurred him on.

He had already presented a paper, *On the general solution of the problem of the polyhedra*, to the Manchester Literary and Philosophical Society at the beginning of 1958. He announced in a letter to Sylvester in January that "the problem of the Polyedra is completely solved in its most general form". But he introduced his paper with a frank admission of the difficulties he had had when generating all the figures involved from a pyramid. The problem was not only that the figure could be derived in different ways from the same pyramid, but that the same figure could be derived from different pyramids.

> And when I have attempted by authority to put the work into the
> hands of some one of these methods, all equally clamorous and
> confident, I have invariably found that, in spite of all I could do to
> prevent it, the operator would persist in carving the same polyedron
> over and over again in different postures, so that it was impossible
> to how many really distinct ones had been generated.

He prepared a paper, written in French, on the enumeration of polyhedra, which he proposed to submit for the Grand Prix in 1861. But, as will be seen below, he became disillusioned and disgruntled with the French over another issue and did not send in his paper. As it happened the Prize was not awarded, though there had been a number of submissions. Meanwhile, Kirkman had sent a preliminary announcement of his work to the Royal Society instead. He then submitted a memoir to the Society; this was in twenty-one lengthy sections, which the referees were unable to cope with. Only the first two sections were published, the rest were deposited in the archives. Norman Biggs describes the first section as dealing with the symmetry of polyhedra and suggests that it implicitly classified the finite groups of isometries in three-dimensional space, but that "the terminology is so complicated that it is difficult to translate any of the theorems into modern terms". For example:

> Effaceables are restored about all like archipoles; the polyarchic
> reticulation laid bare by the removal of these polar summits is
> reduced ... the formulae for polyarchic coronation are given, and
> the result of effacement are enumerated and registered.

Kirkman bombarded George Stokes, who was at the time the Secretary of the Society, with letters, starting with ones that expressed a readiness to improve his text and then turning into angry arguments. The following extracts are typical:

> (14.12.61): I have been unable to discover the slightest value in the
> remarks of your Referee ... In all the work I have not been able to
> find one example of the carelessness of composition with which he

charges me; and I wonder if it be a simple piece of incivility only excusable on the ground of his not understanding the subject, when he says that I have used certain definitions and notions 'from no other purpose than to save the author the trouble of defining them'.

(20.12.61): It is quite plain that your referee contented himself with reading the first page of my book: and his criticism of that page is a nice specimen of schoolmastering. He treats me as a boy answering in Euclid.

(2.1.62): I am willing to believe that the Referee who wrote intended
… what he said to be facetious. His comic talent is very small,
though in Mathematics he is a giant … This book of mine is rather
a large pie to be cooked in England, in which he should not have
had a finger.

He began to feel persecuted and blamed people like Cayley who had
had reservations about Kirkman's intricate style. Moreover, Cayley
published a paper on polyhedra in 1862, which was not much more than a
restatement of Kirkman's early work, and also used some of Kirkman's
terminology. Cayley must have sent Kirkman some account of this for a curt
reply, written on Christmas Day, 1862, included the comment that "your
figure is merely a regular 24-dron trying to look facetious at the expense of
a cube". In the circumstances, it is perhaps surprising to note the postscript:
"If you have any foreign postage stamps, one of my little boys would be
thankful for them – mean defaced only".

Though he continued to work on polyhedra in the last two decades of
his life, he only submitted this later work to the Liverpool Literary and
Philosophical Society. Biggs suggests that Kirkman's obscure papers remain
all that is known about the general classification of polyhedra: "For over a
century the uncompromising tip of the iceberg has successfully deterred
investigation of its submerged portion."

4.2 Solution of equations, many-valued functions and groups

The problem of finding solutions to algebraic equations has always preoccupied mathematicians. The simplest equations that turn up are the ones
that can be expressed in terms of sums of various powers of a single
variable, the so-called polynomial equations. Such equations are named
after the highest power, known as the degree of the equation.

The Babylonians knew how to solve some second degree equations.
These arose from problems involving squares; for example, calculating
dimensions of a square given its area and perimeter. Later, these were called
quadratics, the Latin root indicating the four sides of a square rather than
the degree of the equation. Other equations were given also given Latin
names, namely cubics and quartics for equations of the third and fourth
degrees. By the sixteenth century, mathematicians were able to solve all
equations of degree less than five. The construction of solutions with the
newly developed algebraic notation involved elementary operations such as

adding, multiplying and taking roots. Equations that contained fifth or higher powers of the variable proved to be more difficult. By the nineteenth century, some mathematicians were beginning to suspect that it might not be possible to construct solutions for such equations in general, using only the previous elementary operations.

A fruitful approach to the general problem had been taken by Joseph-Louis Lagrange in 1770, when he analysed the known methods for dealing with quadratic, cubic or quartic equations. He found a common link between these apparently quite disparate methods by showing that they all involved certain functions of the roots. This may be illustrated with quadratic equations: the solution of $x^2+2bx+c=0$ may be reduced to that of a simpler form $y^2=D$, where $y=x+b$ and $D=b^2-c$. This equation has immediate solutions, $y_1=+\sqrt{D}$ and $y_2=-\sqrt{D}$, and the required solutions, x_1 and x_2, of the original equation can be expressed in terms of these. Conversely, the y-solutions can be expressed in terms of the x-solutions: $y_1=(x_1-x_2)/2$ and $y_2=(x_2-x_1)/2$. The second of these has the same form as the first, but with x_1 and x_2 interchanged. This observation can be expressed by saying that the y-solutions are given by a certain function of the x-solutions by permutation of the variables: $y_1 = F(x_1, x_2)$ and $y_2 = F(x_2, x_1)$ where, in this particular case, $F(u,v) = (u-v)/2$

The number of y-solutions, which is the degree of the equation in x is also the number – in this case, two – of different possible values of the function F when the variables are permuted. Lagrange suggested that the clue to the solution of higher degree equations might lie in their corresponding functions. Thus, for quartic equations, he found a function of the four roots, namely $x_1.x_2 + x_3.x_4$, which only had three possible values under the twenty-four possible permutations of the roots. These three values would be the roots of a cubic equation that could be derived from the original equation by means of elementary operations. It was then possible to recover the roots of the original equation from this reduced one.

Lagrange was able to show that this method failed in the case of the fifth degree equation. Although there were functions that yielded a reduced equation, it was more difficult to see how to recover the original roots. It was eventually shown by Niels Abel, in 1824, that the fifth-degree equation could not be solved in elementary terms by any method, and people then began to think in terms of constructing solutions using more complicated functions. Other approaches to the general problem were developed soon after by Evariste Galois, though it was sometime before these became widely known. Meanwhile, people continued to be interested in the question of how functions behave under all possible permutations of the variables. This eventually led to a study of the structure of the permutations themselves and so to the development of the notion of a mathematical *group* – the actual word being first used by Cayley in 1854.

The underlying combinatorial issue may be thought of as an attempt to develop the notion of algebraic symmetry. The word 'symmetry' derives from a Greek word meaning 'common measure' and was used technically in Greek mathematics to mean 'commensurable'. For example, the diagonal and side of a square could not be both expressed as an integral multiple of some common unit, so were said to be *asymmetros* or incommensurable. The word was also used generally to mean harmony, or the quality of being well-proportioned or balanced. It was first used in English in this general sense, and then in the elementary geometrical sense.

By the nineteenth century, the word was beginning to be used in the algebraic contexts: for example, it was applied to expressions like $yz+zx+xy$ which are in some sense 'balanced' in the three letters involved. The search for algebraic symmetries became an important underlying theme in much nineteenth-century mathematics, a theme that was closely linked with the search for the *invariants* in any transformation. As the notion developed, it was then applied in geometrical contexts other than the original bilateral sense; for example, it became possible to think of rotational invariance as a symmetry.

The notion of symmetry was at first invoked in algebra as a metaphor, relying perhaps more on the general sense of harmonious balance. The investigation of how functions behaved under permutations of their variables helped to clarify the notion. The degree of symmetry in a function will be reflected in the number of values it takes when its variables are permuted. Thus a completely symmetric function, such as $x+y+z$, is unchanged by any permutation of its three variables. A completely unsymmetric function, such as $x+2y+3z$, is changed by every permutation and has six different values, corresponding to the six possible permutations. Various possibilities can occur inbetween – for example, $x+y-z$ takes 3 values – but for all cases the number of values can only be a divisor of 6, namely 1, 2, 3 or 6. The corresponding general theorem for functions of n variables is, in effect, what is now called Lagrange's theorem on the possible orders of a subgroup; but this is of course not the way these early results were being interpreted at the time.

Lagrange's condition is necessary but not sufficient: there may not be any function corresponding to a particular divisor. Thus, Paolo Ruffini had shown in 1799 that there was no function of five or more variables with 3 or 4 values. In 1845, Augustin-Louis Cauchy extended this result to up to p values where p was the largest prime number not greater than the number of variables involved. Thus, there could be no functions of six variables with

3, 4 or 5 values. In 1845, Cauchy wrote an enormously long paper – of more than 500 pages! – on the general topic; he referred very handsomely to his younger contemporary, Charles Hermite, who had succeeded in constructing a function of six variables which could take on the permissable number of 6 values.

This brings us back to Sylvester! For the article he wrote in 1861 in which he made some historical remarks about the origins of the schoolgirls problem, was primarily about a different priority claim. The article was called *A Note on the Historical Origin of the unsymmetric Six-valued Function of six letters*. He began by noting that the first construction of such a function was usually assigned to "my illustrious friend" Hermite, ever since Cauchy had expressly done so in 1845. But ...

> I was not at that date in the habit of consulting the *Comptes Rendus*, or I should have at once made the reclamation of priority which I now do, not from any unworthy motive of self-love in so small a matter, but out of regard for historic truth. It is a year or two since I first learned that the origin of the function was usually referred to M. Cauchy or M. Hermite; but although aware that its existence was known to myself long previous to the dates quoted, I did not recollect that I ever communicated it to the world through the medium of the press, and therefore I kept silence on the subject.

Sylvester then explained how he had been looking through some old journals when his eye "chanced to alight" on a footnote to the 1844 paper in which he had introduced the notion of a syntheme (as discussed in section 1.4). He went on to reproduce the two paragraphs of this footnote and to make some further comments. In the footnote he had pointed out how various three-valued functions of the four roots a,b,c,d, of a quartic equation could be constructed from the 'diadic syntheme' (ab, cd). Sylvester had given the example $(a+b).(c+d)$, noting that the roots of the quartic could then be recovered from the cubic equation giving the three values of this function. He then went on to consider six elements a,b,c,d,e,f. In this case, there are fifteen possible diadic synthemes, namely (ab, cd, ef), (ac, be, df), (ad, bf, ce), (ae, bd, cf), (af, bc, de) and so on.

Any one of these would yield various fifteen-valued functions, but these would be no help in solving an original equation of the sixth degree. At first sight, it might be thought possible to group the fifteen synthemes in threes or fours in order to construct five- or three-valued functions that might be of more use. But Sylvester asserted that this would be futile since the synthemes could only be grouped naturally by doubling them up, in which case the duplicate synthemes could be grouped in six groups of five.

Nothing further was said in the 1844 footnote about sixth-degree equations or six-valued functions. But in his 1861 paper, Sylvester wrote out the six sets of synthemes he had referred to in the footnote – one of these consisted of the five synthemes listed above. Each of the six sets he gave could be derived from the other by a permutation of the six letters. Hence any one of them could be said to yield a six-valued function of six elements; for example, the function obtained by taking the product of the sums of the pairs in each of the aforementioned five synthemes.

Sylvester had not specifically mentioned this in 1844 and he would surely have done so if he had drawn this conclusion at the time. His priority claim in this case seems to have some more substance in it than the one about the schoolgirls that he went on to make in the 1861 paper. But the claim does seem rather slender, and it hardly justifies the grandiose way he presented it.

> I conceive that, after this reference, no writer on the subject wishing to specify the function in question would hesitate to call it after my name.

It is again a poignant irony that no-one now refers to the six-valued function of six elements by any name; it has been long since discarded and superceded.

When Kirkman in 1858 read about the Prize Question about polyhedra being set by the Académie des Sciences, he also learned that a previous challenge had proved too difficult and that a different version was to be offered as the Prize Question for 1860, namely

> Quels peuvents être les nombres de valeurs des fonctions bien définie qui contiennent un nombre donné de lettres, et comment peut-on former les fonctions pour lesquelles il existe un nombre donne de valeurs?

This took up the current interest in what were known as many-valued functions and their associated substitution groups. Kirkman had not previously worked in this developing theory, but his combinatorial instincts were aroused and he determined to submit memoirs for this earlier Prize as well as the one on polyhedra which was scheduled for 1861. He published a dozen or so papers over the next few years. The early ones were not much more than a translation of known results into his own terminology – what are now known as left and right cosets, he called derivates and derangements.

He submitted a lengthy memoir to the Académie, as did the French mathematicians, Emile Mathieu and Camille Jordan. The awarding body presented their report in 1861: the three candidates were praised but their work was not considered worthy enough to be awarded the prize. The French mathematicians were highly commended; the verdict on Kirkman's paper was brusque: it was thought that his "notation ingenieuse" could certainly be simplified, that the work was but a rough sketch, and that despite its length it contained nothing new or important.

This rejection angered him and it was this that led him not to submit his work on polyhedra for the next Prize, and to remain very scornful of the Académie. He immediately read his memoir, *On the theory of groups and many-valued functions*, to the Manchester Literary and Philosophical Society and it was published the following year, with a tart comment by Kirkman on the way the prize was not awarded.

> Not the briefest summary was vouchsafed of what the competitors had added to science, although it was confessed that all had contributed results both new and important; and the question, though proposed for the first time for the year 1860, was contrary to the *very frequent* custom of the Academy, withdrawn from competition.

The memoir also included some comments on what he called his *tactical* method. He contrasted this with the "algebra, with its formidable army of congruences and imaginaries" of the French and Italian mathematicians.

Kirkman worked on group theory for about five years, during which he published a dozen or so papers; the last of these referred to a further memoir – on transitive groups – which was never published, possibly because of his continuing criticisms of the Académie. He did return to the subject in 1891 (at the age of 85), when he published a further paper, in which he pointed out that the Prize Question for 1860 had been formulated in terms of many-valued functions rather than groups: "I have most convincing proofs that in the very highest places of European science, this cart before the horse is analytic orthodoxy". This was, he suggested, because the continental mathematicians preferred "pottering behind an orthodox wheelbarrow to doing the work like an Englishman".

Norman Biggs notes that group theory was not taken up in England for thirty years after Kirkman's early papers: "by that time Kirkman's work had been almost completely forgotten".

4.3 Knots

Kirkman's later work on polyhedra also included a discussion in 1881 of a conjecture by Peter Tait that every trivalent polyhedra had a Hamiltonian circuit. This, he wrote, "mocks alike at doubt and proof"; the conjecture was eventually disproved in 1946. The contact with Tait led Kirkman to become interested in the theory of knots that Tait been developing.

One way of classifying a knot is in terms of its crossing number k, namely the number of double points in the simplest plane projection of the knot. There is only one knot with $k=3$ (ignoring mirror reflections), the trefoil or cloverleaf. The only knot with $k=4$ is the figure-eight knot. There are 2,3,7 knots with $k=5,6,7$ respectively. Thereafter the numbers increase dramatically: there are more than twelve thousand knots with k«13. In 1876 Tait had enumerated all knots with 7 or less crossings – these are shown below.

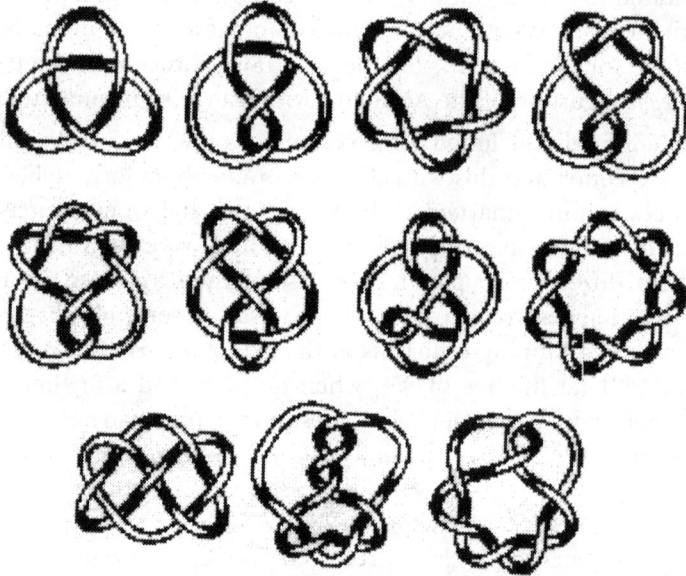

Kirkman collaborated with Tait in extending the list to knots with up to 10 crossings. His first paper on the topic, *The enumeration, description and construction of knots of fewer than ten crossings*, was published in the *Transactions of the Royal Society of Edinburgh* in 1884. The following extract shows he still retained his unique, exuberant style at the age of 78.

> By a knot of n crossings I understand a reticulation of any number of meshes of two or more edges, whose summits, all tesseraces, are each a single crossing, as when you cross your forefingers straight or slightly curved, so as not to link them, and such meshes that every

thread is either seen, when the projection of the knot with its n crossings and no more is drawn in double lines, or conceived by the reader of its course when drawn in a single line, to pass alternately under and over the threads to which it comes at successive crossings.

In this context, knots are always conceived as closed, so that the loose ends of a piece of string that has been knotted in some sort of knot are taken to be then joined up. Knots were then classified in terms of the way they could be decomposed into circuits; a knot that could be traced in one circuit was called *unifilar*, in two circuits *bifilar*, and so on. Kirkman did not include what he called 'solid knots' whose diagrams were projections of polyhedra.

Of solid knots we are not treating. If the apparent dignity of knots so maintains itself as to make a treatise on these n-acra desirable, it will be no difficult thing to show in a future memoir how to enumerate and construct them to any value of n without omission or repetition. The beginner can amuse himself with the regular 8-edron, which is trifilar, or with the unifilar of eight crossings made by drawing a square within a square askew, and filling up with eight triangles.

The unifilar example given by Kirkman is a projection of a square antiprism – a polyhedron with ten sides, two of them squares and the rest triangles. This yields a knot with eight crossings as shown below with alternate crossings-over emphasised.

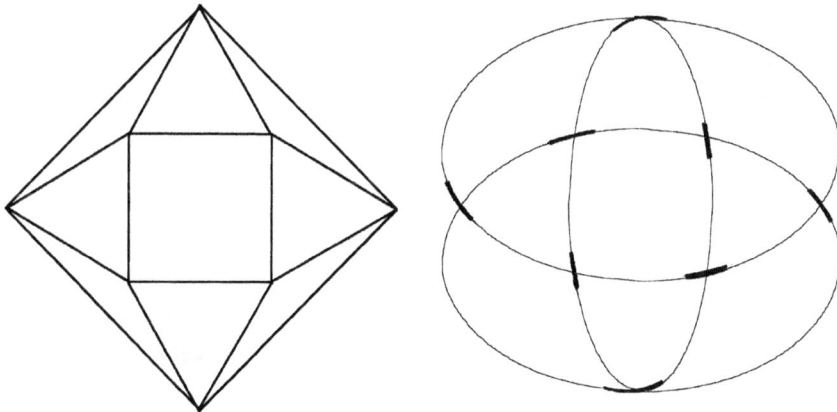

The other example was the projection of a regular octahedron: when this is traced with alternate under-and-over crossings it turns out that there are three independent circuits – in Kirkman's notation, the knot is trifilar – with six crossings. The circuits are known as Borromean rings, after the fifteenth-century Milanese family who adopted them as a part of their coat of arms. The knot has been a symbol of a triple interlinking in various cultures at various times.

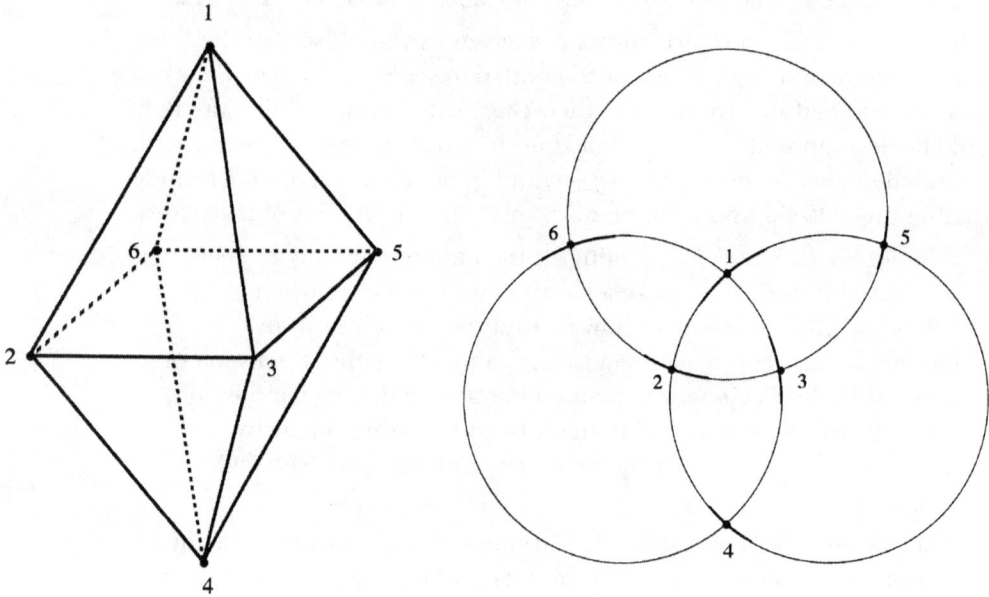

Kirkman assumed that the enumeration of solid knots would easily follow ftom his extensive work on polyhedra. He knew that there were three unifilar knots with six crossings and that these all involved some pairs of points being connected in two ways, so that they contained what Kirkman called *flaps*.

> If a thread a of a knot, after passing under or over a thread b,
> passes over or under b before it meets a third thread c, there is a
> linkage of two crossovers and a flap between them. This flap is the
> eyelet seen between two links of a slack chain as it lies on the table:
> it is a 2-gon, a mesh of two edges and of two crossings.

Of course, solid knots, which were projections of polyhedra, did not have any flaps. Kirkman enumerated other knots which did and claimed to have found 634 unifilar knots with ten crossing, that were not projections of polyhedra.

Kirkman continued for a year or two to investigate the subject. And as usual, he did so in his own inimitable jargon; for example, two points involved in a certain operation were referred to as "the crossing P and the crease-kiss R, or, as a more learned man would say, through the decussation

P and the plicatorial osculation R". He was not always clear about the way that knots with apparently different crossings could in fact sometimes be topologically equivalent. Few of his contemporaries pursued his ideas – and even Tait found some of Kirkman's headier calculations too tedious to follow. But some of his approach was been taken up in later developments of knot theory.

4.4 Mathematical puzzles

The *Educational Times*, a monthly supplement of the *Times* newspaper, was first published in 1847, the year in which Kirkman wrote his first mathematical paper. This always included a number of mathematical problems for readers, who were invited to submit solutions. These became well-known and attracted the attention of professional, as well as amateur, mathematicians, and so led to a semi-annual publication called *Mathematical Questions and Solutions from the Educational Times*. First published in 1863 and continued for more than fifty years, this was once described by William Clifford as contributing more to mathematical research than any other European journal.

As might be expected, Kirkman was a frequent contributor and continued to submit mathematical problems in his later years. One example was Question 7933, proposed in 1886 when he was eighty, and written in his customary doggerel style. This described how thirteen people were threatened by Puck, the fairy: "Before another year can fly / Some shall sicken and one shall die". But then Puck relented and commanded that they should meet four at a time at every new moon for thirteen months:

> New fours, for thirteen months, be told
> Their penitent watch on the hill to hold;
> But no two twice, of the banned thirteen,
> May see together the moonlit scene.

The solution may be interpreted as a finite projective plane with thirteen points lying in fours on thirteen lines, and is a particular case – namely a (4,2) system on 13 elements – of a theorem in his 1850 paper that there is a $(y,2)$ system covering all pairs of y^2-y+1 elements, for prime $(y-1)$.

In another contribution, he posed a problem whose solution was a (5,2) system on 25 elements. In this case Old King Cole commands his 25 fiddlers to perform in groups of five for six days with any pair performing together once and only once.

> 'Full oft ye have had your fiddle's fling
> for your own fun over the wine;
> And now,' quoth Cole, the merry old king,
> 'Ye shall have it again for mine.
> My realm prepares for a week of joy
> At the coming of age of a princely boy –
> of the grand six day's procession in square,
> In all your splendours dressed,
> Filling the city with music rare
> From fiddlers, five abreast.'

This is again a particular case of a more general theorem from his 1850 paper, namely that there is a $(y,2)$ system covering all pairs when y is prime and the number of elements is a power of this prime.

Kirkman discussed some other multiple systems that had been proposed in 1869 by another contributor. He confirmed the existence of a (4,3) system of 16 elements, this being a particular case of a theorem he had published in 1852 that there was always a (4,3) system for any number of elements that was a power of 2. But he thought that a proposed (5,4) system on 15 elements was unlikely; such a system was not in fact shown to be impossible until 1972.

Not all his contributions were about the multiple systems that he had investigated so thoroughly. For instance, a lengthy doggerel had an African king visiting London with his fifteen wives ("every one a Venus vast of twenty stone") and 15 pairs of named lords and ladies. These names were such that each involved six vowels (including 'y'), made up of four vowels with two of them repeated. Some complicated conditions determined how many dinners they could attend. The only solution submitted was that of the proposer. It turned out that the required permutations of the vowels in the names involved a particular six-valued function of six letters. "Six dinners, therefore, at the most, were eaten."

Another problem, set in 1868, involved the solution of a quintic equation. He then claimed, in an article in the *Philosophical Magazine*, to have solved the quintic by a method that could be applied to any algebraic equation. But he did later that year admit he had made a mistake: "I hope the scientific reader will pardon the nonsense of the latter part of my preceding communication, and I wish that he may live to confess, like myself, that his mathematical powers are the worse for wear."

A rather simpler puzzle was sent to his friend, Alexander Macfarlane, who mentioned it in a chapter on Kirkman in his book on some nineteenth-century mathematicians.

Baby Tom of Baby Hugh
The nephew is and uncle too;
In how many ways can this be true?

Kirkman continued in his eighties to send problems to the *Educational Times*. In 1892, the year in which he retired at the age of 86, he submitted a complicated combinatorial problem involving ten women students from Girton College, Cambridge.

In ninety-two and ninety-three,
Twelve balls of ours we fix to be:
Ten Girton maids, Urania's pets,
We bid in unnamed quintuplets,
To each to come, five left at home,
While dance a dozen different sets,
The beauties don't deny, but all
Cry: 'Most unmathematical!
Shall we, for random invitations,
Descend into Combinations?'
....

A further verse introduces the requirement that each student attends the same number (and so in this case six) of balls, But it turns out that the restriction to a dozen balls does not permit a balanced system. In one solution that was printed later that year, each student attends six balls but not all pairs attend the same number of balls. Kirkman then followed this up with an amazing variation that constrained the numbers of times that one student A attends balls with the eight others, apart from one specified exception B, so that these numbers were perfect squares! It is perhaps not surprising that there was no solution submitted to this problem.

Kirkman would have appreciated a puzzle arising from a remark by his contemporary Augustus de Morgan who pointed out that he was aged N in the year N^2? As also was Kirkman! Some other mathematicians sharing this feature would be Thomas Harriott (born 1560), Philippe de la Hire (born 1640) and Stefan Banach (born 1892). It may also be worth noting that Harry Potter will be 45 in 2025, but it is not yet clear whether he will be a mathematician.

5

Combinatorial developments

5.1 Tactical configurations, Moore systems, and block designs

In his first papers on multiple systems, Kirkman had only considered $(y,2)$ systems of multiples that covered all pairs. The original problem in the *Lady's and Gentleman's Diary* in 1844 had asked for the construction of more general (y,z) systems in which y-sets (that is, sets of y elements) cover all z-sets. In a further combinatorial paper, *Theorems on combinations*, published in 1853, Kirkman made his only step in this difficult general direction by constructing a $(4,3)$ system, namely a system of quadruples covering all triples, with any number of elements that was a power of 2. He presented this as a problem about a 'crocodile' of schoolgirls, 2^n of them walking in fours with every three of them walking together once and only once.

General (y,z) systems were looked at from another point of view by an American mathematician, Eliakim Moore, in a paper, *Tactical memoranda*, published in 1896, the year after Kirkman died. Moore here used a word that had been coined by Sylvester and often used by Kirkman – 'combinatorial' would now be the most common choice. He noted that the configurations extensively studied by geometers in the nineteenth century had some tactical characteristics in common with the multiple systems studied by Kirkman (and now often called Moore systems). The geometrical configurations were, of course, particular $(y,2)$ systems, which have already been discussed in a previous section.

Moore established some necessary conditions on the number x of elements of a (y,z) system. In the case of Kirkman's $(4,3)$ systems these conditions meant that x must be of the form $6k+2$ or $6k+4$ (which include Kirkman's powers of two), and this was proved to be sufficient by Haim Hanani in 1963. It is known that there are unique $(4,3)$ systems with 8 or 10 elements, four distinct systems with 14 elements and at least eight with 16 elements.

Few other general results about (y,z) systems are known. In fact only a few systems have been constructed for $z>3$. Moore observed that if all the y-sets not containing a certain element A were suppressed and this element is deleted from the remaining sets, there is left a system of $(y-1)$-sets that cover all $(z-1)$-sets, namely a $(y-1,z-1)$ system. This means, for instance, that the $(6,5)$ and $(8,5)$ systems that were later discovered also implied the existence of $(5,4)$ and $(7,4)$ systems. The rich complexity of the $(8,5)$ system, discovered in 1938, is illustrated by the fact that its 759 octuples cover every one of its 24 elements 253 times, every pair 77 times, every triple 21 times, every quadruple 5 times and every quintuple once. This covering of successive multiples from single elements to quintuples can be described as a (253, 77, 21, 5, 1) *balance*. A recent, sophisticated application of this particular system has been the derivation of a particularly close packing of spheres in 24-dimensional space!

The repeated covering of submultiples in the (8,5) system, from a single element up to quadruples, is a characteristic feature of multiple systems. Kirkman's 1853 paper on combinations was mainly concerned with such repetitions. He found systems with a regularly repeated cover of all pairs; for example, his paper began with an example of a (4,2) system of quadruples of 7 elements with every pair occurring in three and only three quadruples. Kirkman constructed some more general examples of (3,2) systems of triples of $12k+3$ elements that covered all pairs $6k+2$ or $6k+5$ times. He also quoted the configuration of the 28 bitangents of a quartic curve, which can be arranged in quadruples, whose eight points of contact lie on a conic to form a complete system of 315 conics of which 45 pass through the points of contact of each bitangent, and 5 pass through the four points of contact of each pair of bitangents. Thus, the 28 bitangents form a (4,2) system covering all pairs 5 times. Kirkman generalised this to the construction of (4,2) systems of quadruples of $12k+4$ elements covering all pairs $2k+1$ times.

Though piecemeal in its approach, Kirkman's generalisation to repeated covers was an original and important step. Systems of multiples that cover all pairs a regular number of times were intensively studied and widely applied in the twentieth century. As long as the cover is not repeated, there will be some geometrical representation for these systems, but when the cover is repeated the structures remain purely 'tactical', though as is common in mathematics some geometrical thinking is still pervasive.

These general systems were introduced into the design of biological experiments in the 1930s. When Ronald Fisher was the chief statistician at the government agricultural centre at Rothampstead, he had observed that biological experiments were not producing clear-cut conclusions and he began to study the scientific principles of experiment design. (The latin-square arrangement of trees mentioned in section 2.2 was designed by Fisher.) In many cases, it was not possible to test all the elements of a variety at a time and it became important to choose smaller (incomplete) samples, but to randomise (balance) the distribution as much as possible. At first, the schedules for doing this were known as incomplete randomised block arrangements.

An example, introduced by Frank Yates, who was Fisher's successor at Rothampstead, was a (3,2) system of triples that covered all pairs twice. There were 6 elements forming 10 triples in which each single element occurred 5 times. The covering of single letters and pairs could be described as a 5,2 balance. One representation could be the following three-letter words:

   YEA   PER   YET   PAY   PRY   TAP   ARE   RAT   TRY   PET

An example of a specific application of such experimental schedules could be an experiment in 1942 to compare the poisonous effect on a certain species of aphid of six glycinotride compounds with that of a standard nicotine spray. Only three sprays could be tested on one day and the reactions of the aphids were expected to vary from day to day. The schedule, or design, adopted was for 7 blocks of three sprays over seven days. Each spray was tested three times in all, and every possible pair of sprays were tested together once. This called for a (3,2) system of triples covering all of the 7 elements three times and all pairs once – a Kirkman triple system that may be represented by the following three-letter words:

   ADO   ORE   BAR   BOY   YEA   BED   DRY

In general, the universal set of elements was called a *variety* and the multiples were called *blocks*. The interaction of blocks and pairs gave what statisticians called a balanced incomplete block design. Statisticians usually refer to a block design as a (v,b,k,r,λ)-*configuration*, though the parameters can be given in other orders. This may be described as a $(k,2)$ system with v elements and b of the k-sets (blocks) that cover each element r times and each pair λ times. The parameters are redundant and, as is shown below, the system can be sufficiently characterised by specifying just v, k and λ.

Moore had started his 1896 paper by considering some very general tactical configurations. He then defined a particular (two-dimensional) tactical configuration in terms of a binary relation between two sets such that every element of one set was related to the same number of elements of the other set. When the relation was particularised so that the relation was repeated a certain number of times the configuration was said to be balanced. This would be (using the notation that has been used throughout) a (y,z) system with the y-sets covering all z-sets a number, say λ, of times, and this may be described as a (y,z,λ) tactical configuration.

In this case, the z-sets containing any one of the x elements may be counted in two ways. If any element A lies in Y of the y-sets the number of z-sets containing A will be Y times the number of $(z-1)$-sets in the part of a y-set not containing A, namely $C(y-1,z-1)$. But it will also be λ times the number of possible $(z-1)$-sets, namely $C(x-1,z-1)$, so that

   $Y = \lambda.C(x-1,z-1) / C(y-1,z-1) = \lambda.(x-1) ....(x-z+1) / (y-1) ....(y-z+1)$

and this must be an integer. Moreover if there are X of the y-sets, then counting the number of elements in two ways gives $X.y = x.Y$ so that

$$X = \lambda.x(x1) \dots (x-z+1) \,/\, y(y-1) \dots (y-z+1)$$

and this must also be an integer. These provide necessary conditions on the parameters involved in the description of any tactical configuration. As an example, consider the representation of a (4,2,3) configuration with 8 elements represented by the 8 vertices of a unit cube. Then there will be $X=14$ quadruples and $Y=7$ of these will contain any one vertex. Here the quadruples may be represented by the six faces of the cube, the six diagonal planes and the two regular tetrahedra formed by the vertices, while each vertex will lie in three faces, three diagonal planes and one tetrahedron.

The Moore-Kirkman (y,z) systems can be described as $(y,z,1)$ tactical configurations. Particular examples would be the $(y,2,1)$ Kirkman systems, including the (3,2,1) Kirkman-Steiner triples. With the extended notation, a balanced incomplete block design may be described as a $(y,2,\lambda)$ tactical configuration. With x elements, this is a system of $\lambda.x(x-1)/y(y-1)$ blocks, or y-sets, that covers all elements $\lambda.(x-1)/(y-1)$ times and all pairs λ times, where the necessary conditions on the parameters mean that the two algebraic expressions must yield integers. As is the case with the particular designs with $\lambda=1$ there have been various attempts to ascertain whether these conditions are sufficient, that is, whether there *are* designs with parameters that satisfy the conditions. Generalising his results for multiple systems, Hanani has shown that this is so when $y=3$, 4, or 5 for all λ except that there is no (5,2,2) design with 15 elements. More generally, Robin Wilson has shown that for a wide range of values of y and λ, a block design always exists if the variety (that is, the number of elements) is sufficiently large. He also showed that if this is so for $\lambda=1$, then it is for all λ.

Most of the combinatorial structures directly developed from Kirkman's work, such as schoolgirl and other triple systems, multiple systems, finite projective planes and planar difference sets, are particular examples of balanced block designs with $\lambda=1$, and so of general tactical configurations.

5.2 Real roots, anallagmatic pavements and Hadamard matrices

It took some time for mathematicians to formulate the notion that an algebraic equation of the n-th degree in one variable had n roots – and even longer to prove this satisfactorily. To assert, for instance, that a quadratic equation always has two roots is to have some understanding and confidence in 'imaginary' numbers. The development of the so-called fundamental theorem of algebra is part of the history of complex numbers.

But during this development there was a continuing interest in the problem of determining the number of *real* roots of a given equation, and of distinguishing which of these were positive and which negative. Many mathematicians contributed to the problem by finding upper bounds for the number of positive roots, and so also for negative roots by changing the sign of the variable in the equation.

The earliest, and simplest, bound was stated without proof by Descartes in an appendix to his *Discours de la Méthode* of 1637. His 'rule of signs' asserts that there are no more positive roots than changes of sign in the coefficients of the equation taken in ascending or descending order. Thus, for example, the equation $x^3-2x^2+x+1=0$, which has two changes of sign in its sequence of coefficients $(+ - + +)$, cannot have more than two positive roots. Descartes' rule of signs was not in fact proved until the eighteenth century, though improvements had been found by then. Thus, Newton gave improved bounds, but with a more complicated rule, in his *Arithmetica Universalis*, published in 1707, but based on lectures given thirty years earlier.

The problem was to some extent settled in 1835 when a Swiss mathematician, Charles Sturm, gave an algorithm for determining the exact number of real roots of an algebraic equation in one variable. The calculation of Sturm's algorithm was a standard pinnacle of higher mathematics in late nineteenth-century textbooks.

As it happened, Sylvester became interested in Newton's rule setting an upper bound to the number of real root, and he proved it as part of a more general result in 1865. This was published as the first paper of the *Proceedings of the London Mathematical Society*, which was founded in that year. In his customary way, Sylvester tried to place his work on roots of an equation into a wider context. He created a particularly rich texture of cross-references in 1867, when he published a paper in the *Philosophical Magazine* with the remarkable title: *Thoughts on inverse orthogonal matrices, simultaneous sign-succession and tessellated pavements in two or more colours, with applications to Newton's rule, ornamental tile-work and the theory of numbers.*

Without going into much detail, the main threads of this paper can be described from the observation that Sylvester's version of a rule of signs for the enumeration of the real roots of an algebraic equation led him to consider square arrays of positive and negative signs with the curious property that any two rows of the array agree in half their places and disagree in the other half.

Such arrays can clearly be presented as a tessellation of squares coloured in the appropriate way with two colours. Translated into the language of matrices that had been developed over the previous decade – principally by

Cayley – the arrays became square matrices whose elements are $+1$ or -1. The 'equal agreement' property of the rows of signs ensures that the product of the matrix and its transpose is the identity matrix multiplied by the number of elements in a row. Such matrices – called *orthogonal matrices* by Charles Hermite in 1855 – are essentially unchanged by inversion.

Sylvester coined the adjective *anallagmatic* (from Greek for 'unchanging') when presenting the arrays in the guise of a colouring problem on a chessboard for readers of the *Educational Times* for November 1867:

> Show that it is possible to paint the squares [of a chessboard] in a manner such that on comparing any two rows or columns, four of the squares opposite one another shall have like, and the remaining unlike, colours

Sylvester asked also for constructions of anallagmatic pavements with 2^n squares in a row, instead of 8. An easy way of constructing the pavements he asked for is to compound pavements of order 2. In general a pavement of order m and one of order n can be compounded to give a pavement of order $m.n$. To do this, squares in one colour in the first pavement are replaced by the whole second pavement and the other squares are replaced by the second pavement with al its colours reversed. This means that a pavement cane be constructed by replacing a single square A by a 2×2 square B, with elements as shown where the squares of A' are coloured in the opposite colour to those of A, and then repeating this process.

$$B = \begin{matrix} A & A \\ A & A' \end{matrix} \qquad C = \begin{matrix} B & B \\ B & B' \end{matrix} \qquad D = \begin{matrix} C & C \\ C & C' \end{matrix} \qquad \ldots\ldots$$

Further problems on the theme were set by Sylvester in 1868 and 1869 and anallagmatic pavements became a standard topic in recreational mathematics.

The problem re-appeared in its matrix form in 1893 when Jean Hadamard was investigating the possible values of determinants with elements in the range -1 to $+1$. He found that $n \times n$ determinants could not exceed $n^{n/2}$. He also showed that this upper bound was only attained in the case of precisely those matrices considered by Sylvester, namely orthogonal matrices whose elements are $+1$ or -1. Such matrices have now become known as *Hadamard matrices*. It turns out that the order (namely the number of elements in any row or column) of a Hadamard matrix must be 1, 2 or a multiple of 4. This does not mean that there is such a matrix for every multiple of 4, though only a few exceptions have been found.

Sylvester's pavements indicate that Hadamard matrices provide a form of combinatorial design, so that it is not surprising that they were considered from this point of view in the nineteen-thirties when such designs were being developed. The general situation may be illustrated by a particular case. Take a 12×12 Hadamard matrix and remove the row and the column whose elements are all ones. The resulting 11×11 array, shown below with O,X for +1, −1, has five Os in each row and any two rows have two pairs of Os in the same column. This yields a combinatorial structure for 11 elements arranged in quintuples that cover all pairs twice and only twice. In terms of the notion of balance introduced in the previous section, this could be described as a quintuple system with balance (11,5,2) and so, in contemporary terms, a particular form of a symmetric balanced incomplete block design.

```
X O X O O O X X X O X
X X O X O O O X X X O
O X X O X O O O X X X
X O X X O X O O O X X
X X O X X O X O O O X
X X X O X X O X O O O
O X X X O X X O X O O
O O X X X O X X O X O
O O O X X X O X X O X
X O O O X X X O X X O
O X O O O X X X O X X
```

Hadamard matrix as anallagmatic pavement

1	2	3	4	5	6	7	8	9	10	11
.	O	.	O	O	O	.	.	.	O	.
.	.	O	.	O	O	O	.	.	.	O
O	.	.	O	.	O	O	O	.	.	.
.	O	.	.	O	.	O	O	O	.	.
.	.	O	.	.	O	.	O	O	O	.
.	.	.	O	.	.	O	.	O	O	O
O	.	.	.	O	.	.	O	.	O	O
O	O	.	.	.	O	.	.	O	.	O
O	O	O	.	.	.	O	.	.	O	.
.	O	O	O	.	.	.	O	.	.	O
O	.	O	O	O	.	.	.	O	.	.

(15,11,2) system of quintuples of 11 elements covering each pair twice and only twice

There was renewed interest in Hadamard matrices after the second world-war, as it was felt that their defining property provided an optimum discrimination between rows and so a good basis for error-correcting codes. Computer searches established further matrices leaving only one unresolved case (of order 188) up to 200. It is still conjectured, but not proved, that Hadamard matrices exist for *all* multiples of 4.

5.3 Mathematical form

One early attempt to systematise the notion of a general configuration can be found in the work of Alfred Kempe. Like Cayley and Sylvester, he had been a Cambridge graduate (in fact, a student of Cayley's) and had then become a lawyer. Unlike them, he remained a lawyer all his life, but he continued his interest in mathematics and made a number of important contributions, for which he became a Fellow of the Royal Society.

Kempe was inspired by a lecture given by Sylvester on linkages, which were drawing devices consisting of hinged rods which could be used to trace curves. His book on linkages published in 1877 included the remarkable theorem that there is always some linkage to trace any curve represented by an algebraic equation. Two years later, he wrote a paper on a conjecture, current at the time, that four colours would suffice to colour the regions of any planar map. Unfortunately his proof was faulty, as was pointed out eleven years later by Percy Heawood. That the conjecture was in fact correct was not proved until a computer verification in 1976.

In 1885, Kempe wrote *A memoir introductory to a general theory of mathematical form*. This was a very abstract account which identified certain common components of various branches of mathematics.

> The object of the memoir is the treatment of the 'necessary matter' of exact or mathematical thought as a connected whole; the separation of its essential elements from the accidental clothing – algebraic, geometrical, logical, &c. – in which they are usually presented for consideration; and the indication of that to which the infinite variety which these elements exhibit is due.

He characterised the elements of what he called a *system* by suggesting that, as far as reasoning about them was concerned, they are, whatever their real nature – "as points, lines, statements, relationships, arrangements, intervals or periods of time, algebraic expression" – always dealt with as distinct entities or *units*. Some pairs, or other multiples of such units could be distinguished from each other, while others could not. Every collection of units were said to have a definite *form* that arose from "the way in which the distinguished and un-distinguished units, pairs, triads, &c., are distributed throughout the collection".

In the present context, Kempe's work may be seen as an early attempt to systemise the notion of a configuration and to formulate some overall view of the combinatorial ideas that were in the air in his time. His ideas were developed further in a talk which he gave to the London Mathematical Society on the occasion of his retirement as its President in 1894. In discussing the nature of mathematics, he referred briefly to various dictionary definitions such as "the science of number and magnitude" and to the more ambitious definitions of some of his contemporaries, such as De Morgan's "space and time ... form the subject-matter of mathematics", or Benjamin Peirce's characterisation of mathematics as "the science which draws necessary conclusions". But he found these unsatisfactory on the grounds that "such statements could hardly have furthered the progress of mathematics at any period of its history". He preferred to frame a definition that "might reasonably be expected to open out new fields, and to indicate lines of connexions between mathematical subjects regarded as fundamentally distinct".

Kempe drew attention to a characteristic feature of mathematics at the time, namely the predominant importance of theories which dealt with discontinuous quantities. Echoing his previous emphasis on the distinct units of a system he suggested that mathematics involved a number of "individual conceptions":

> To enumerate such examples as quantities, points, curves, spaces, algebras, equations, letters, arrangements, substitutions, rotations, differentials, numbers, statements, and quaternions, conveys but a faint idea of the variety exhibited. The various individuals under consideration are not, of course, jumbled together in a mere confused heap, but bear relations to each other.

He pointed out that various branches of mathematics were beginning to be seen as interconnected, and he suggested that this should also be the case for the classical distinction between pure and applied mathematics. The mathematical properties "evolved out of the inner consciousness", were essentially the same as those which were "obtained as the result of experience and observation". In general, a correspondence could often be found between any two distinct branches of mathematics. "To each individual thing, relation or property, which is of importance in the one case, there corresponds respectively an individual thing, relation or property, in the other." It was this correspondence which Kempe called *form*.

The like and unlike individuals and pluralities which are contained in any greater plurality must be distributed in some way through the whole body of individuals composing that greater plurality, and the way in which this distribution is effected gives to the latter a characteristic form which may be the same, or may differ, in two pluralities of the same number of individuals.

One example began with ten elements of a regular tetrahedron, namely the four vertices and the six edges. The elements (or individuals) of each of the two sets (or pluralities) are essentially similar, but in the case of a square this aspect of form is quite different. Thus, although the four vertices are also essentially similar, in the case of the six edges – namely joins of vertices – the two diagonals differ in nature from the four sides. Another comparison was with the vertices and sides of a regular pentagon. Here there are again ten elements but they are distributed in different ways; in particular, in the way they group into pairs. Thus in the case of the tetrahedron there are 6 pairs of adjacent points, but in the pentagon there are 5 such pairs. The tetrahedron has 12 pairs of opposite edges, the pentagon has 5 such pairs – and so on. "Without entering upon the consideration of other peculiarities which might be pointed out, it is obvious that the two systems have marked difference of form."

Kempe gave another example in which two widely differing sorts of system had the same form. One system consisted of the four vertices A,B,C,D of a regular tetrahedron and the twenty-four triangles obtained by joining the vertices to the midpoints of opposite sides. There is one and only one triangle – shaded in the figure here – such that no side of it passes through A, the longest side passes through B, the shortest side passes through C, and the remaining side passes through D. This may be denoted by a symbol (ABCD), and the other triangles will then be denoted by the permutations of the vertices. The other system consisted of four variables w,x,y,z and the twenty-four permutations of a linear function $aw+bx+cy+dz$ of these variables; these permutations may be characterised in usual notation by the symbol $(wxyz)$. Noting that the four variables in each case are essentially similar. Kempe observed that "the correspondence between the two complete systems of twenty-eight individuals becomes pretty obvious".

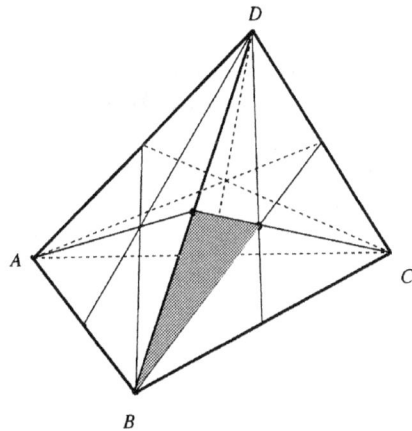

Where two systems had the same form there would be a precise correspondence between their mathematical properties. In particular, this provided an explanation of the importance of algebra. "In any algebraic representation of mathematical facts, we merely strip of all the mathematically irrelevant clothing ... and re-clothe the essential element – form – in a dress more convenient for the purpose of its investigation." Thus in the last example the four elements clothed as points are replaced by four other elements, now clothed as variables, and the twenty-four quadruples clothed as triangles are replaced by twenty-four other elements, now clothed as permutations.

Kempe ended his talk with a provisional definition of mathematics: Mathematics is the science by which we investigate those characteristics of any subject-matter of thought which are due to the conception that it consists of a number of differing and non-differing individuals and pluralities.

Kempe's first wife died in 1893 and he married again four years later. He was knighted in 1912 for his legal services to various government departments, but his health deteriorated at about this time and he was to die of pneumonia in 1922 at the age of 73. There were some moving obituaries which testified the wide regard in which he was held. One account of his work as a lawyer mentioned that "the clarity of his mind made it difficult for him to argue a rotten point". Another referred to the way "his humility of mind and antipathy to anything like self-advertisement read a continual lesson to the ambitious".

5.4 The fifteen schoolgirls

It may be recalled that in his first paper on combinations, Kirkman considered a problem which was a particular case of the Prize Question posed by Woolhouse three years before. In the notation that has been consistently used in the previous sections, the general problem was one of finding the greatest number of combinations in which x elements could be arranged as a (y,z) system. Kirkman considered the particular case $(3,2)$, and showed that such a system could not always be complete in the sense that it included all possible pairs. Since pairs can be selected from x elements in $x(x-1)/2$ ways, but that these can be asssociated with another element in 3 ways, a complete system will have $x(x-1)/6$ triples. Kirkman showed that this could only be possible if and only if x was of the form $6k+1$ or $6k+3$.

It was this result that led him three years later to propose the particular case when $x=15$ in terms of the schoolgirls, where the completeness of the system meant that all the possible 35 triples could be covered by a daily crocodile of 5 triples for – conveniently – a week. The scheduling of schoolgirls on a daily walk over a week seemed but a trivial puzzle to interest readers of an annual publication. But it caught the attention of a number of mathematicians at the time and continued to provide problems that were not in fact settled until much later. Kirkman had realised that the particular case of the 15 schoolgirls had meant that they could indeed be scheduled to have $(15-1)/2$ different walks and that each girl walked each day. But it was not clear when this could be extended to x schoolgirls scheduled for $(x-1)/2$ walks. Certainly in this case x could not be of the form $6k+1$ for it had to be divisible by 3 for every girl to be able to walk each day. Hence for the general form of the problem, x could only be of the form $6k+3$. Kirkman was aware that it had not yet been proved that this was a sufficient condition in this particular case.

Both Cayley and Sylvester were attracted by the problem. Cayley considered the case $x=7$ as well as $x=15$ and linked these with extensions of Hamilton's quaternions to 8- and 16-dimensional numbers. He also noted a further extension by Sylvester: "to make the schoogirls walk every week in the quarter so that each triple may walk together", so that there would be different weekly schedules for 13 weeks.

In his 1850 paper on Woolhouse's Prize Question, Kirkman acknowledged Sylvester's "charming extension" and gave his own extensions for the 84 triples when $x=9$. He also noted that the Rev. James Mease had sent him a solution of the basic question for $x=27$ with a general theorem when x was any power of 3. Kirkman provided a generalisation of his own for the case when x was five times a power of 3, as well some theorems about multiple $(y,2)$ systems which have been mentioned in a previous chapter.

Kirkman had established conditions on the number of elements forming a complete triple system. But it was not clear whether there was always an automatic way of generating such a system. Hence the early interest in offering actual schedules for various permissable numbers of elements. Thus, in 1852, another cleric, Robert Anstice, constructed a (3,2) system for a number of elements of the form $6k+3$ where $3k+1$ is prime, and so for $x=15$, 27, 39, ... His thirteen page paper in the *Cambridge and Dublin Mathematics Journal* ends with a note that he could add more "but too much space has been already devoted to such a trifle"; he did add more in the same journal a year later. Kirkman was obviously delighted at the interest being shown. In a letter to the *Philosophical Magazine* he notes that "the puzzle of the fifteen young ladies now takes rank as a case of a mathematical problem", and pays tribute to the work of Anstice and to another contributor, William Spottiswoode, an Oxford don and later a President of the Royal Society.

The schoolgirls problem attracted the attention of an American mathematician, Benjamin Peirce, the father of the philosopher Charles Peirce. In 1860, he extended Anstice's result for numbers of the form $6k+3$, where k is even or one more than a multiple of four, and so now including $x = 33,57, 81...$ He also discussed a general $(y,2)$ system with y^2 elements; for example, the scheduling of 16 schoolgirls in quadruples over 5 days. This was again discussed by Woolhouse in his paper in the following year (mentioned in section 1.3) in which he took issue with Sylvester about his claim to have originated the schoolgirls problem. This $(4,2)$ system is now invoked in the scheduling of speedway tournaments in which 16 riders race in heats of four at a time, with each pair racing together once and only once.

By the end of the nineteenth century, the general problem of arranging x elements in a $(3,2)$ system in $(x-1)/2$ blocks had been settled for values of x of the form $6k+1$ up to 100. There had also been a number of proposals for ways of generating the required schedule for various cases. For example, in the figure below the five triangles shown provide a possible schedule for the first day's walk by 15 schoolgirls. The schedules for the rest of the week may now be read off by rotating the triangles two steps round the the central vertex.

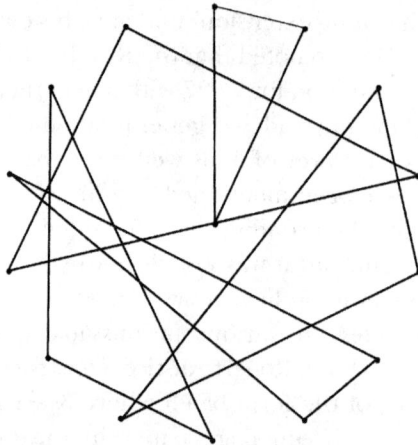

Later examples included that of a 3-dimensional finite geometry with 15 points on 35 lines each having three points – the required solutions are taken from 7 sets of 5 skew lines. The figure below is a representation of the 15 points on a tetrahedron OABC. Each of the four faces represent a 7-point geometry, so that for instance the circle $XB'C'$ represents one of the seven lines. One of the possible schedules for the 15 schoolgirls is indicated by the 5 emphasised skew lines (one of which is that circle).

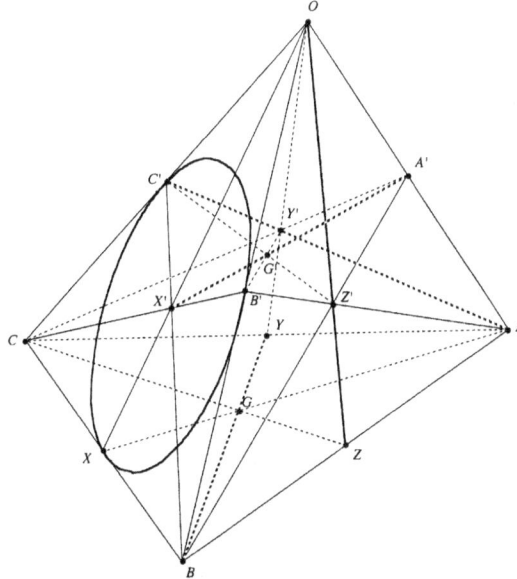

Meanwhile in the twentieth century, Kirkman's triple were being seen as very particular cases of block designs and these became the main focus of combinatorial attention. But the basic problem of proving the sufficiency of Kirkman's condition remained, and this was eventually proved by Dwijendra Ray-Chaudhuri and Richard Wilson in 1971. To arrange x elements in $(x-1)/2$ blocks of $x/3$ triples covering each pair and with each element occurring once in every block, it is both necessary and sufficient that x be of the form $6k+3$. It is interesting to note that the subsidiary theorems leading to this result included another extension of Anstice's result, namely to when $3k+1$ is any power of a prime, so now adding $x=49, 97\ldots$

A final twist to the tale concerns Sylvester's observation that the scheduling of the 15 schoolgirls could be repeated over 13 weeks, thus including all the 455 possible triples. In the general case with $6k+3$ elements the schedule for a 'week' of $3k+1$ 'days' needs to be repeated over $6k+1$ 'weeks'. Some constructions have been found for particular cases, but no general method has been found. Kirkman had mistakenly thought he had found a construction in the case of the 15 schoolgirls; one was not found until a computer-generated solution was given in 1974 by Ralph Denniston. It was this that was to inspire the composer, Tom Johnson, who set the triples to music in *Kirkman's Ladies*, written in 2005:

Each lady/note appears once in the daily phrases of five chords, each pair of ladies walks together once a week, and by the end of the 13 weeks, all 455 possible trios of women have passed by, as have the 455 chords that represent them ... I particularly like to imagine my three-note chords played by three flutes, or as a harp solo, though an interpretation with three oboes, three strings or one vibraphone might also be just fine, and since the music is really only notes and numbers, I would not want to prevent anyone from playing it on a piano or with some other instruments. We should leave the chords in this register, though, and always keep the ladies clean and pretty. It seems safe to assume that the sun is always shining – otherwise they would not be taking walks.

Epilogue

Hail the spirit able to unite!
For we truly live our lives in symbol,
and with tiny paces move our nimble
clocks beside our real day and night.

Still we somehow act in true relation,
we that find ourselves we know not where.
Distant station feels for distant station –
what seemed empty space could bear

...

It is relatively easy to give an account of Kirkman's mathematical work. He published more than seventy papers in various journals, he corresponded with many mathematicians, and he was elected a fellow of the Royal Society. Although he would at one time have liked to have had an academic post he remained an amateur; but he was respected by many of his professional contemporaries. If much of his work was not fully appreciated in his time, it can be seen now – with hindsight – to have foreshadowed many important later mathematical developments.

It is not so easy to capture something of the personality of the creator of all this original mathematics. It seems that he was much respected in his small parish, where his duties would not have been too onerous and would have left him with lots of time to pursue his mathematical and philosophical interests. Despite his occasional grumbles, he seems to have established warm relationships with some of his correspondents. In a letter to Sylvester, dated 2 Jan. 1858, mainly about a possible teaching post, he introduced a rare personal note: "I regret to learn from your letter that one so formed as you are for connubial bliss is taking so little pain to attain it, and is still endeavouring to reconcile himself to the fate of a batchelor". And the letter ended: "Mrs K joins me in wishing you many happy new years". In the following month, he wrote again with a worry that he might have offended Cayley, but added a warm (underlined) invitation to Sylvester: "I will certainly call upon you when I come to London and you will at any time be <u>a most welcome guest here</u>".

Another side of him is occasionally revealed in other correspondence. Some of his letters to George Stokes, in which he complains about the way his work on polyhedra had been treated by the Royal Society, have already been quoted. He also wrote to Stokes on other matters and here he often betrayed a slightly obsessive pedantry, usually tinged with his own particular brand of humour. In one letter in 1862, he comments on the mathematical philosophy of John Stuart Mill:

A) Mr Mill tells us that he has gone a long way in mathematics.

B) Mr Mill never went so far as the end of his nose: Mr Mill denies the existence of a mathematical point, which verily exists at the end of his nose, having no position and no parts.

From A and B it follows, if induction is worth anything, that Mr Mill wears a nose of fortuitous dimensions.

Kirkman also corresponded with the physicist James Maxwell, who had been interested in knot theory. In one letter he criticised a book – by their mutual friend Peter Tait and William Thompson (later Lord Kelvin) – which had referred to matter as if it was a personified agent:

In every English grammar – in the penny manuals of the lowest standards – it will have to be stated that while *he can*, *she can*, *they* (living agents) *can*, is English. *It can* (active) is neither sense nor English. *It can* is the phrase of the savage trembling before the fetish.

Such brief quotations merely hint at what must have been a fascinating, complex and highly original mind. We do not know much about his domestic life and can only guess what it must have been like in the relatively untypical, middle-class Victorian household, where the father would be at home for much of the week, often working at mathematics in his study, while the mother would be looking after the house and the seven children (probably with some domestic help). He is known to have been interested in teaching children (though his book on mnemonic aids to learning mathematics did not seem to stem from very much actual experience) and he spent some time training young choristers in his parish. We know nothing about his actual relationships with parishioners or with his family, though it is known that he was particularly distraught when his fourth son, Frederic, died at the age of thirty.

What is even more tantalising is that we know so little about his own childhood, or indeed about his first twenty years. He was an only son and had several sisters (the usual brief biographical accounts never specify the actual number of sisters). He gained a place at the local grammar school where he impressed his teachers as being a likely candidate for a Cambridge scholarship. A teacher and the local vicar both urged that the boy should stay at school and even offered some financial help. But his father was adamant and insisted on his son leaving school when he was fourteen in order to come and work in his office.

It is known that Kirkman dutifully carried out mundane clerical work for his father for the next nine years. It also recorded that he privately continued to study whenever he could. Inevitably, we can now only guess at what the experience must have meant for the talented youth. Nine years! What was his future going to be? ... Suddenly, at the age of 23, he left home and enrolled as a student at Trinity College, Dublin, supporting himself by giving private tuition. His subsequent career was briefly described in an earlier section (1.1).

The bald facts are astounding enough: one can again only guess at what lies behind them. Even more tantalising is the lack of any account of his earliest years. What was it like for him in a household of women and a dominating father? What was his relationship with his mother? Is there any significance in the fact that his wife had the same name, Eliza(beth)? Where was he in the sequence of sibling births? What did he feel about his father – in childhood, in his adolescence? What might have been the determining factors in his early upbringing that could help to explain his later intellectual interests? What led to the dramatic move from being a clerk in his father's office to being a university student? What led him to his particular mathematical interests? Was his passion for collecting different combinations and enumerating them a symbolic way of sorting out his feelings, of controlling them in some sublimated way? Was this also at work in his theological and philosophical beliefs, where he seemed particularly concerned to refute views which did not fit in properly with his own?

How did he cope with his successes and with his disappointments? ... What sort of man was he? What, in Rilke's words above, were his real day and night?

Henri Poincaré once famously asked "How is it that there are so many minds that are incapable of understanding mathematics?" The question still baffles mathematics teachers and various answers continue to be proposed; it may be recalled that for Kirkman the problem was an overemphasis in teaching on the eye as opposed to ear. There is always, however, the disturbing possibility that the supposedly incapable are merely choosing what to attend to, what to get involved in, what to resist. It is perhaps the converse question that should be being proposed: "How is it that so many minds can become so preoccupied with mathematics?" How and why did Kirkman become a mathematician? ...

The questions mount up and have to remain unanswered. Meanwhile, those fifteen schoolgirls are still taking their daily walks. Though it may be recalled that fifteen schoolgirls were burned to death in a fire at their school in Mecca on 11 March 2002, when Saudi Arabian police prevented then from leaving the building because they were not wearing headscarves and had no male relatives there to receive them. Furthermore, it seems that a choir of fifteen schoolgirls sang at a memorial service, on 18 March 2003, to the peace activist, Rachel Corrie, who was crushed to death by an Israeli bulldozer as she tried to prevent the destruction of Palestinian homes. As Kirkman observed in a paper read to the Liverpool Literary and Philosophical Society in 1877: "Wonderful and fertile in crime and cruelty are and have been the strifes and contradictions of human thought, in the states, in the churches, and in the schools".

...

purest tension. Harmony of forces!
Do not just our limited resources
keep all interference from your flow?
Does the farmer, anxiously arranging,
ever reach to where the seed is changing
into summer? Does not Earth bestow?

(Rainer Maria Rilke, Sonnets to Orpheus, 1.XII, trans. J. B. Leishmann)

William Tahta, *Fifteen things*, 2006

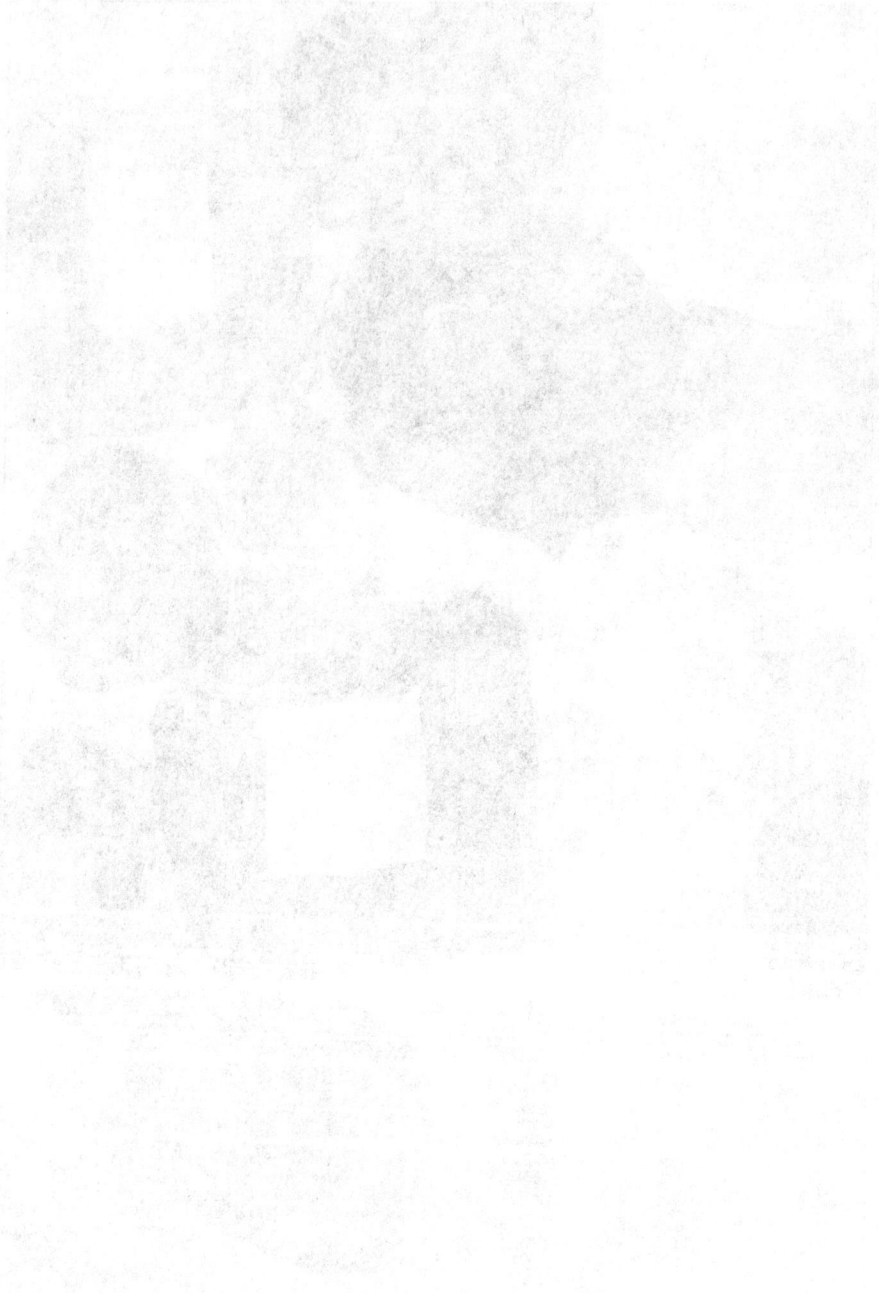

www.ingramcontent.com/pod-product-compliance
Lightning Source LLC
Chambersburg PA
CBHW051223200326
41519CB00025B/7231